CW00705097

Chemical Bonding in Crystals and Their Properties

Chemical Bonding in Crystals and Their Properties

Special Issue Editors

Anna V. Vologzhanina
Yulia V. Nelyubina

MDPI • Basel • Beijing • Wuhan • Barcelona • Belgrade • Manchester • Tokyo • Cluj • Tianjin

Special Issue Editors
Anna V. Vologzhanina
Russian Academy of Sciences
Russian

Yulia V. Nelyubina
Russian Academy of Sciences
Russian

Editorial Office
MDPI
St. Alban-Anlage 66
4052 Basel, Switzerland

This is a reprint of articles from the Special Issue published online in the open access journal *Crystals* (ISSN 2073-4352) (available at: https://www.mdpi.com/journal/crystals/special_issues/chemical_bonding).

For citation purposes, cite each article independently as indicated on the article page online and as indicated below:

LastName, A.A.; LastName, B.B.; LastName, C.C. Article Title. *Journal Name* **Year**, *Article Number*, Page Range.

ISBN 978-3-03936-170-0 (Hbk)
ISBN 978-3-03936-171-7 (PDF)

Contents

About the Special Issue Editors

Anna V. Vologzhanina (PhD) is Senior Researcher at the X-ray Structure Centre of A.N. Nesmeyanov Institute of Organoelement Compounds of Russian Academy of Sciences (INEOS RAS). She graduated from Samara State University in 2003, and received her PhD degree in 2006 from the same university. Since 2013, she has been managing the research activity of the Russian National Affiliated Centre of the Cambridge Crystallographic Data Centre. In 2014, she was awarded with the National Fellowship (Russian Federation) L'Oréal-UNESCO "For Women in Science". She is a co-author of more than 200 papers in the field of crystal chemistry of actinides, lanthanides, and transition metals as well as charge density studies and CSD analysis.

Yulia V. Nelyubina (PhD, Dr. Habil.) is Head of Centre for Molecular Composition Studies at A.N. Nesmeyanov Institute of Organoelement Compounds of Russian Academy of Sciences (INEOS RAS). In 2008, she graduated from the Higher Chemical College of Russian Academy of Sciences, and obtained her PhD degree and Habilitation in Physical Chemistry from INEOS RAS in 2009 and 2018, respectively. In 2012, she received the National Fellowship (Russian Federation) L'Oréal-UNESCO "For Women in Science", and in 2013, was awarded the Academia Europaea Prize. She has co-authored more than 200 papers on X-ray diffraction and charge density studies of various inorganic, organic, and organometallic compounds prior to focusing her scientific interests on magnetic molecular materials for switching, sensing, and other applications.

Editorial

Special Issue Editorial: Chemical Bonding in Crystals and Their Properties

Anna V. Vologzhanina * and Yulia V. Nelyubina

A. N. Nesmeyanov Institute of Organoelement Compounds, RAS. 28 Vavilova str., 119991 Moscow, Russia; unelya@ineos.ac.ru
* Correspondence: vologzhanina@mail.ru

Received: 9 March 2020; Accepted: 11 March 2020; Published: 12 March 2020

Relations between physicochemical properties of chemical compounds exploited in many modern applications (including optical, magnetic, electrical, mechanical, and others) and interatomic interactions that operate in their crystals are the key to the successful design of new crystalline materials, in which X-ray crystallography has proved to be an invaluable tool. In addition to the advanced approaches in charge-density analysis that provide insights into the nature of chemical bonding, the information collected over the years by this technique and stored in huge databases has a tremendous use in drug design and other areas of material science.

This Special Issue covers a diverse range of 'structure–property' and 'composition–structure' relations identified through X-ray diffraction. Two reviews [1,2] and five articles [3–7] were submitted and published.

In reference [1], possible interconnections between crystal properties and molecular and crystal structures were summarized. This paper clearly demonstrates how the knowledge of molecular geometry and intermolecular interactions, of bonding preferences for some motifs, synthons and tectons extracted from the Cambridge Structural Database can be used for material chemistry, crystal engineering, pharmaceutical, and agrochemical research. Numerous examples of polymorphism rationalization, co-crystal design, control over crystal morphology, rationalization of mechanical and sorption properties, and studies of hydration/dehydration mechanisms were described.

The review [2] on chemical bonding in crystals of low-melting organoelement compounds allowed for the identification of a linear relation between the molecular volume or a Hirshfeld surface area and the energy of the crystal lattice for compounds with similar types of predominant intermolecular interactions. It was demonstrated that these compounds are typically involved in weak- and medium-strength interactions while strong bonding, if any, is responsible for the formation of isolated molecular associates.

As the analysis of weak intermolecular interactions requires highly accurate experimental or computational data, the authors of [3,5] used periodic density functional theory (DFT) calculations to study the role of F...F interactions in fluorinated tosylates and of halogen and chalcogen bonding in thiazolo[2,3–b][1,3]thiazinium triiodides, respectively. Such an approach when combined with the quantum theory of "Atoms-in-Molecules", electron localization function, noncovalent interactions method, or other partitioning schemes provides insight into weak interatomic interactions and even quantifies their strength. In particular, in reference [3], an almost linear dependence was uncovered between the contribution of interactions involving fluorine atoms to the lattice energy and the amount of fluorine atoms, although its increase does not lead to crystal packing stabilization. On the other hand, numerous noncovalent interactions of triiodides were attributed to a stronger I–I bond within the triiodide anion, acting as a stabilizing factor and providing a comparatively higher thermal stability and iodine retention in the melt [5].

Among other experimental techniques, Raman spectroscopy was found to be useful for understanding the bonding features of the triiodide anion [5]; however, weak intermolecular interactions can also be identified by other spectroscopic tools. For example, a study [4] of a series of salts of boron cluster anions with protonated organic bases demonstrated that dihydrogen bonds have characteristic absorption bands in the FT-IR spectra of solids, which can therefore be used to recognize these bonds even in the absence of crystallographic data [4].

Unlike the other papers in this Issue, reference [6] focuses on intermolecular interactions and supramolecular associates found in crystals of RNA. An analysis of H-bond connected sextuples of RNA bases collected in the Protein Data Bank and relative occurrences of the sextuples allowed the authors of [6] to classify some of them as a novel RNA tertiary motif.

In a comparative study of strong hydrogen bonds and weak interactions in racemic and enantiopure thiophosphorylated thioureas [7], a new synthetic pathway was suggested to control the chirality of their Ni(II) complexes at both the molecular and supramolecular levels.

In summary, this Special Issue covers very different aspects of structure–property relations identified by X-ray diffraction and complementary techniques (from conventional IR and Raman spectroscopies to cutting-edge quantum chemical calculations) and their application in crystal engineering and material science.

Author Contributions: A.V.V.: writing—original draft preparation, Y.V.N.: review and editing. All authors have read and agreed to the published version of the manuscript.

Funding: This work was supported by Ministry of Science and Higher Education of the Russian Federation.

Acknowledgments: The authors are grateful to all the other authors who contributed to this Issue, and to the Editorial Office of Crystals for their efforts to make our contributions open to the scientific community.

Conflicts of Interest: Authors declare no conflict of interest.

References

1. Vologzhanina, A.V. Intermolecular Interaction in Functional Crystalline Materials: From Data to Knowledge. *Crystals* **2019**, *9*, 478. [CrossRef]
2. Volodin, A.D.; Korlyukov, A.A.; Smol'yakov, A.F. Organoelement Compounds Crystallized in Situ: Weak Intermolecular Interactions and Lattice Energies. *Crystals* **2020**, *10*, 15. [CrossRef]
3. Arkhipov, D.E.; Lyubeshkin, A.V.; Volodin, A.D.; Korlyukov, A.A. Molecular Structures Polymorphism the Role of F...F Interactions in Crystal Packing of Fluorinated Tosylates. *Crystals* **2019**, *9*, 242. [CrossRef]
4. Avdeeva, V.V.; Vologzhanina, A.V.; Malinina, E.A.; Kuznetsov, N.T. Dihydrogen Bonds in Salts of Boron Cluster Anions $[B_nH_n]^{2-}$ with Protonated Heterocyclic Organic Bases. *Crystals* **2019**, *9*, 330. [CrossRef]
5. Yushina, I.; Tarasova, N.; Kim, D.; Sharutin, V.; Bartashevich, E. Noncovalent Bonds, Spectral and Thermal Properties of Substituted Thiazolo[2,3-b][1,3]thiazinium Triiodides. *Crystals* **2019**, *9*, 506. [CrossRef]
6. Hamdani, H.Y.; Firdaus-Raih, M. Identification of Structural Motifs Using Networks of Hydrogen-Bonded Base Interactions in RNA Crystallographic Structures. *Crystals* **2019**, *9*, 550. [CrossRef]
7. Kataeva, O.; Metlushka, K.; Yamaleeva, Z.; Ivshin, K.; Zinnatullin, R.; Nikitina, K.; Sadkova, D.; Badeeva, E.; Sinyashin, O.; Alfonsov, V. Chirality Control in Crystalline Ni(II) Complexes of Thiophosphorylated Thioureas. *Crystals* **2019**, *9*, 606. [CrossRef]

Article

Chirality Control in Crystalline Ni(II) Complexes of Thiophosphorylated Thioureas

Olga Kataeva [1,*], Kirill Metlushka [1], Zilya Yamaleeva [1], Kamil Ivshin [1,2], Ruzal Zinnatullin [1], Kristina Nikitina [1], Dilyara Sadkova [1], Elena Badeeva [1], Oleg Sinyashin [1] and Vladimir Alfonsov [1]

[1]   Arbuzov Institute of Organic and Physical Chemistry, FRC Kazan Scientific Center, Russian Academy of Sciences, Arbuzov str. 8, 420088 Kazan, Russia; metlushka@mail.ru (K.M.); zika0527@mail.ru (Z.Y.); kamil.ivshin@yandex.ru (K.I.); zruzal94@mail.ru (R.Z.); kristina-nik-25@mail.ru (K.N.); Dilyara-Sadkova@yandex.ru (D.S.); ybadeev.61@mail.ru (E.B.); oleg@iopc.ru (O.S.); alfonsov@yandex.ru (V.A.)

[2]   A.M. Butlerov Chemistry Institute, Kazan Federal University, Kremlevskaya street 18, 420008 Kazan, Russia

[*]   Correspondence: olga-kataeva@yandex.ru

Received: 29 October 2019; Accepted: 19 November 2019; Published: 20 November 2019

Abstract: Chirality control over the formation of Ni(II) complexes with chiral thiophosphorylated thioureas was achieved via breaking the symmetry of nickel coordination geometry by the introduction of the pyridine ligand, while centrosymmetric *meso*-complexes are formed from racemic ligands in case of square-planar nickel coordination. Centrosymmetric heterochiral arrangement is observed in crystals of ligands themselves through N–H···S hydrogen bonds in intermolecular dimers. Molecular homochirality in tetragonal pyramidal complexes is further transferred to supramolecular homochiral arrangement via key–lock steric interactions.

Keywords: chiral thiophosphorylated thioureas; chirality control; nickel(II) complexes; X-ray single crystal diffraction

---

1. Introduction

The comparison of intermolecular interactions in the crystals of enantiopure and racemic compounds is of primary importance to address the questions of chiral recognition, interplay between molecular and supramolecular chirality, bioactivity of chiral drugs, self-sorting and finally, the origin of homochirality [1–7]. Chiral recognition or preferential interactions between the enantiomers of the same chirality is difficult to achieve for conformationally flexible molecules, which possess multiple functional groups able to participate in a variety of molecular interactions. This might include hydrogen bonding, π-stacking, steric interactions, metal coordination, etc. At the same time, multiple possible intermolecular interactions upon certain conditions may provide not only discreet homochiral species, but chiral recognition on different levels: molecular level, formation of homochiral 1D-supramolecular chains, 2D-homochiral nets and finally, chiral resolution of racemic species. The strength and the directionality of molecular interactions leading to stable rigid supramolecular aggregates is the decisive factor for chiral recognition [8], e.g., the formation of centrosymmetric hydrogen-bonded dimers of chiral carbonic acids is the prevailing supramolecular synthon composed of enantiomers of opposite chirality [9,10]. This strong interaction prevents the formation of homochiral supramolecular species.

Strange enough, metal coordination is rarely used to achieve chiral recognition, though coordination bonds are comparatively strong and directional. Recently [11,12], we have demonstrated chiral recognition in the crystals of Ni(II) complexes with chiral 1-(1-phenyl)ethyl-3-(*O*,*O*-diethylthiophosphoryl)thioureas on different levels, including the formation of homochiral complexes and conglomerate crystals. Chiral thiophosphorylated thioureas are ideal compounds to study the processes of chiral recognition. They possess several hydrogen bond acceptors and donors, are able to coordinate metal ions, have conformationally flexible terminal groups that provide multiple modes of crystal packing depending on the crystal growth conditions. More important, one can introduce a variety of chiral auxiliaries of different volume and topology.

Herewith, we present the data on new chiral thiophosphorylated thioureas and their nickel(II) complexes in racemic and enantiopure form, addressing the stereochemical aspects of the molecular and supramolecular arrangement.

2. Materials and Methods

2.1. Chemistry

2.1.1. General

^{1}H NMR spectra were recorded on an AVANCE-400 (Bruker, Karlsruhe, Germany) instrument with the working frequency of 399.93 MHz relative to the signals of residual protons of deuterated solvents (CDCl$_3$, C$_6$D$_6$), ^{31}P NMR spectra were obtained on an AVANCE-400 (Bruker, Karlsruhe, Germany) instrument with the working frequency of 161.90 MHz relative to the external standard (85% H$_3$PO$_4$). IR spectra have been registered using a Tensor 27 Fourier spectrometer (Bruker, Karlsruhe, Germany) in the 400–4000 cm^{-1} range (optical resolution 4 cm^{-1}). The samples were prepared as KBr pellets. The ESI MS measurements were performed using an AmazonX ion trap mass spectrometer (Bruker, Karlsruhe, Germany) in positive mode in the mass range of 70–3000. The capillary voltage was −3500 V, nitrogen drying gas −10 L·min^{-1}, desolvation temperature −250 °C. The sample was dissolved in MeCN or DMF to a concentration of 10^{-6} g·L^{-1}. Data processing was performed by DataAnalysis 4.0 SP4 software (Bruker, version 4.0, Karlsruhe, Germany). Optical rotations were determined on a Perkin Elmer (Model 341) polarimeter at 20 °C. Melting points were measured on a BOETIUS melting point microscope.

All chemicals were purchased from Sigma-Aldrich (Moscow, Russia) and used without further purification.

2.1.2. Syntheses

(±)-1-(1,2,3,4-Tetrahydronaphthalen-1-yl)-3-(O,O-diethyl thiophosphoryl)thiourea ((±)-1): *O*,*O*-diethyl thiophosphoryl isothiocyanate (1.9 g; 9 mmol) in acetonitrile (4 mL) was added dropwise to the solution of 1,2,3,4-tetrahydro-1-naphthylamine (1.32 g; 9 mmol) in acetonitrile (8 mL) under stirring. Resulting mixture was stirred at r.t. for 1 day under argon. After that the solvent was evaporated and the obtained viscous oil was recrystallized from the mixture of cyclohexane and ethyl acetate (10:1). The resulting precipitate was filtered off, washed with a small amount of cyclohexane and dried in vacuo to give **(±)-1**. Yield: 2.5 g (77.8%); m.p. 97–99 °C; IR (KBr): ν (cm^{-1}) 3234 (NH), 1541, 1487 (NCS), 1017 (C-O-P), 614 (P=S); ^{1}H NMR (400 MHz, CDCl$_3$): δ (ppm) 1.25, 1.32 (2t, $^3J_{HH}$ = 7.1 Hz, 6H, C$\underline{H}_3$CH$_2$OP), 1.80–2.01 (m, 3H, C$\underline{H}_2$ $_{THNaph}$), 2.15–2.23 (m, 1H, C$\underline{H}_2$ $_{THNaph}$), 2.76–2.91 (m, 2H, C$\underline{H}_2$ $_{THNaph}$), 4.09–4.19 (m, 4H, CH$_3$C$\underline{H}_2$OP), 5.64–5.69 (m, 1H, C$\underline{H}_{THNaph}$), 7.03 (d, $^2J_{PH}$ = 12.0 Hz, 1H, N$\underline{H}$P), 7.12–7.36 (m, 4H, C$\underline{H}_{THNaph}$), 7.90 (d, $^3J_{HH}$ = 7.8 Hz, 1H, N$\underline{H}$C(S)); ^{31}P NMR (400 MHz, CDCl$_3$): δ$_P$ (ppm) 55.97; ESI$^+$-MS (CH$_3$CN): *m/z* 359.1 [M + H]$^+$, 229.1 [M + 2H-{C$_{10}$H$_{11}$}]$^+$; Elemental analysis calcd (%) for C$_{15}$H$_{23}$N$_2$O$_2$PS$_2$: C 50.26, H 6.47, N 7.81, P 8.64, S 17.89; found (%): C 50.34, H 6.54, N 7.71, P 8.41, S 17.60.

Single crystals, suitable for X-ray diffraction analysis, were obtained by slow evaporation of the mother liquor after precipitate filtration.

(R)-1-(1,2,3,4-Tetrahydronaphthalen-1-yl)-3-(O,O-diethyl   thiophosphoryl)thiourea   ((R)-1): preparation method is the same as for **(±)-1** using (R)-1,2,3,4-tetrahydro-1-naphthylamine as the initial amine. Viscous oil after solvent evaporation was dissolved in the mixture of cyclohexane and hexane (20:1) and kept at 5 °C for one week. The resulting precipitate was filtered off, washed with a small amount of hexane and dried in vacuo to give **(R)-1**. Yield: 2.66 g (82.7%); m.p. 90–91 °C; $[\alpha]_D^{20}$ = +37.1 (c 1.0, CHCl$_3$); IR (KBr): ν (cm^{-1}) 3235 (NH), 1544, 1486 (NCS), 1017 (C-O-P), 613 (P = S); ^{1}H NMR (400 MHz, CDCl$_3$): δ (ppm) 1.25, 1.32 (2t, $^3J_{HH}$ = 7.1 Hz, 6H, C$\underline{H}_3$CH$_2$OP), 1.80–2.00 (m, 3H, C$\underline{H}_2$ $_{THNaph}$), 2.14–2.23 (m, 1H, C$\underline{H}_2$ $_{THNaph}$), 2.76–2.91 (m, 2H, C$\underline{H}_2$ $_{THNaph}$), 4.09–4.21 (m, 4H, CH$_3$C$\underline{H}_2$OP), 5.64–5.69 (m, 1H, C$\underline{H}_{THNaph}$), 7.03 (d, $^2J_{PH}$ = 11.9 Hz, 1H, N$\underline{H}$P), 7.12–7.36 (m, 4H, C$\underline{H}_{THNaph}$), 7.90 (d, $^3J_{HH}$ = 7.8 Hz, 1H, N$\underline{H}$C(S)); ^{31}P NMR (400 MHz, CDCl$_3$): δ_P (ppm) 55.97; ESI$^+$-MS (CH$_3$CN): m/z 359.1 [M + H]$^+$, 229.1 [M + 2H-{C$_{10}$H$_{11}$}]$^+$; Elemental analysis calcd (%) for C$_{15}$H$_{23}$N$_2$O$_2$PS$_2$: C 50.26, H 6.47, N 7.81, P 8.64, S 17.89; found (%): C 50.35, H 6.66, N 7.66, P 8.55, S 18.17.

Single crystals, suitable for X-ray diffraction analysis, were obtained by slow evaporation of the mother liquor after precipitate filtration.

(S)-1-(1,2,3,4-Tetrahydronaphthalen-1-yl)-3-(O,O-diethyl   thiophosphoryl)thiourea   ((S)-1): preparation method is the same as for **(±)-1** using (S)-1,2,3,4-tetrahydro-1-naphthylamine as the initial amine. Thiourea **(S)-1** was isolated by the same crystallization procedure as for **(R)-1**. Yield: 2.71 g (84.3%); m.p. 89–91 °C; $[\alpha]_D^{20}$ = −36.8 (c 1.0, CHCl$_3$); IR (KBr): ν (cm^{-1}) 3235 (NH), 1543, 1486 (NCS), 1017 (C-O-P), 613 (P=S); ^{1}H NMR (400 MHz, CDCl$_3$): δ (ppm) 1.25, 1.32 (2t, $^3J_{HH}$ = 7.1 Hz, 6H, C$\underline{H}_3$CH$_2$OP), 1.81–2.01 (m, 3H, C$\underline{H}_2$ $_{THNaph}$), 2.14–2.23 (m, 1H, C$\underline{H}_2$ $_{THNaph}$), 2.76–2.91 (m, 2H, C$\underline{H}_2$ $_{THNaph}$), 4.08–4.19 (m, 4H, CH$_3$C$\underline{H}_2$OP), 5.64–5.69 (m, 1H, C$\underline{H}_{THNaph}$), 7.02 (d, $^2J_{PH}$ = 11.9 Hz, 1H, N$\underline{H}$P), 7.12–7.36 (m, 4H, C$\underline{H}_{THNaph}$), 7.91 (d, $^3J_{HH}$ = 7.9 Hz, 1H, N$\underline{H}$C(S)); ^{31}P NMR (400 MHz, CDCl$_3$): δ_P (ppm) 55.98; ESI$^+$-MS (CH$_3$CN): m/z: 359.1 [M + H]$^+$, 229.1 [M + 2H-{C$_{10}$H$_{11}$}]$^+$; Elemental analysis calcd (%) for C$_{15}$H$_{23}$N$_2$O$_2$PS$_2$: C 50.26, H 6.47, N 7.81, P 8.64, S 17.89; found (%): C 50.42, H 6.69, N 8.02, P 8.43, S 17.69.

(meso)-NiL$_2$-type Complex ((meso)-2): N-thiophosphorylated thiourea **(±)-1** (0.5 g, 1.4 mmol) and potassium hydroxide (0.117 g, 2.1 mmol) were dissolved in methanol (10 mL). The resulting mixture was stirred during 10 min, and after that a solution of Ni(II) chloride hexahydrate (0.199 g, 0.84 mmol) in methanol (5 mL) was added to it. The reaction mixture was stirred at room temperature for a further 24 h. After that, the solvent was evaporated, the resulting solid was dissolved in dichloromethane (50 mL) and extracted by water (2 × 15 mL). The organic layer was separated and dried with anhydrous Na$_2$SO$_4$. Drying agent was filtered off, and the solvent was evaporated. The resulting solid was dissolved in the mixture of chloroform and hexane (3:5). During 2 weeks of slow evaporation of the mother liquor, crystals were formed, which were filtered off and dried in vacuo to give **(meso)-2**. Yield: 0.32 g (59.3%); m.p. 186–187 °C; IR (KBr): ν (cm^{-1}) 3168 (NH), 1562 (NCS), 1041, 1022 (C-O-P), 627 (P = S); ^{1}H NMR (400 MHz, C$_6$D$_6$): δ (ppm) 1.17, 1.22 (2t, $^3J_{HH}$ = 6.9 Hz, 6H, C$\underline{H}_3$CH$_2$OP), 1.26–1.35 (m, 1H, C$\underline{H}_2$ $_{THNaph}$), 1.47–1.64 (m, 2H, C$\underline{H}_2$ $_{THNaph}$), 1.69–1.79 (m, 1H, C$\underline{H}_2$ $_{THNaph}$), 2.17–2.39 (m, 2H, C$\underline{H}_2$ $_{THNaph}$), 3.89–4.22 (m, 4H, CH$_3$C$\underline{H}_2$OP), 5.09–5.17 (m, 1H, C$\underline{H}_{THNaph}$), 6.76–7.05 (m, 3H, C$\underline{H}_{THNaph}$), 7.47–7.51 (m, 1H, C$\underline{H}_{THNaph}$), 9.94 (br.s, 1H, N$\underline{H}$C(S)); ^{31}P NMR (400 MHz, C$_6$D$_6$): δ_P (ppm) 58.13; ESI$^+$-MS (DMF): m/z 773.2 [M + H]$^+$, 359.1 [M + 2H-Ni-L]$^+$; Elemental analysis calcd (%) for C$_{30}$H$_{44}$N$_4$NiO$_4$P$_2$S$_4$: C 46.58, H 5.73, N 7.24, Ni 7.59, P 8.01, S 16.58; found (%): C 46.65, H 5.53, N 7.11, Ni 7.30, P 7.80, S 16.42.

(R,R)-NiL₂-type Complex ((R,R)-2): preparation method is the same as for ***(meso)*-2** using ***(R)*-1** as the initial thiourea. Recrystallization was carried out from the mixture of chloroform and hexane (2:5). During 2 weeks of slow evaporation of the mother liquor, crystals were formed, which were filtered off and dried in vacuo to give ***(R,R)*-2**. Yield: 0.28 g (51.9%); m.p. 136–138 °C; $[\alpha]_D^{20}$ = +245 (c 0.3, C_6H_6); IR (KBr): ν (cm⁻¹) 3183 (NH), 1544 (NCS), 1033, 1014 (C-O-P), 620 (P = S); ¹H NMR (400 MHz, C_6D_6): δ (ppm) 1.17, 1.22 (2t, $^3J_{HH}$ = 7.0 Hz, 6H, C$\underline{H}_3$CH₂OP), 1.26–1.36 (m, 1H, C$\underline{H}_2$ $_{THNaph}$), 1.46–1.65 (m, 2H, C$\underline{H}_2$ $_{THNaph}$), 1.68–1.78 (m, 1H, C$\underline{H}_2$ $_{THNaph}$), 2.18–2.40 (m, 2H, C$\underline{H}_2$ $_{THNaph}$), 3.90–4.23 (m, 4H, CH₃C$\underline{H}_2$OP), 5.10–5.16 (m, 1H, C$\underline{H}$$_{THNaph}$), 6.76–7.04 (m, 3H, C$\underline{H}$$_{THNaph}$), 7.48–7.51 (m, 1H, C$\underline{H}$$_{THNaph}$), 9.95 (br.s, 1H, N$\underline{H}$C(S)); ³¹P NMR (400 MHz, C_6D_6): δ$_P$ (ppm) 58.40; ESI⁺-MS (DMF): *m/z* 773.2 [M + H]⁺, 359.1 [M + 2H-Ni-L]⁺; Elemental analysis calcd (%) for $C_{30}H_{44}N_4NiO_4P_2S_4$: C 46.58, H 5.73, N 7.24, Ni 7.59, P 8.01, S 16.58; found (%): C 46.70, H 5.95, N 6.97, Ni 7.34, P 7.84, S 16.83.

(S,S)-NiL₂-type Complex ((S,S)-2): preparation method is the same as for ***(meso)*-2** using ***(S)*-1** as the initial thiourea. Complex ***(S,S)*-2** was isolated by the same crystallization procedure as for ***(R,R)*-2**. Yield: 0.29 g (53.7%); m.p. 135–137 °C; $[\alpha]_D^{20}$ = −243 (c 0.3, C_6H_6); IR (KBr): ν (cm⁻¹) 3185 (NH), 1544 (NCS), 1033, 1013 (C-O-P), 620 (P=S); ¹H NMR (400 MHz, C_6D_6): δ (ppm) 1.17, 1.22 (2t, $^3J_{HH}$ = 7.0 Hz, 6H, C$\underline{H}_3$CH₂OP), 1.26–1.36 (m, 1H, C$\underline{H}_2$ $_{THNaph}$), 1.47–1.66 (m, 2H, C$\underline{H}_2$ $_{THNaph}$), 1.69–1.78 (m, 1H, C$\underline{H}_2$ $_{THNaph}$), 2.18–2.40 (m, 2H, C$\underline{H}_2$ $_{THNaph}$), 3.88–4.22 (m, 4H, CH₃C$\underline{H}_2$OP), 5.09–5.15 (m, 1H, C$\underline{H}$$_{THNaph}$), 6.76–7.04 (m, 3H, C$\underline{H}$$_{THNaph}$), 7.48–7.51 (m, 1H, C$\underline{H}$$_{THNaph}$), 9.97 (br.s, 1H, N$\underline{H}$C(S)); ³¹P NMR (400 MHz, C_6D_6): δ$_P$ (ppm) 58.43; ESI⁺-MS (DMF): *m/z* 773.2 [M + H]⁺, 359.1 [M + 2H-Ni-L]⁺; Elemental analysis calcd (%) for $C_{30}H_{44}N_4NiO_4P_2S_4$: C 46.58, H 5.73, N 7.24, Ni 7.59, P 8.01, S 16.58; found (%): C 46.81, H 5.93, N 7.04, Ni 7.42, P 8.04, S 16.79.

(rac)-NiL₂·Py-type Complex ((R,R/S,S)-3): Ni(II) acetate tetrahydrate (0.138 g, 0.55 mmol) was dissolved in a mixture of pyridine (0.176 g, 2.2 mmol) and methanol (8 mL, 16 mL, 32 mL). After that, the solution of racemic N-thiophosphorylated thiourea (±)-1 (0.4 g, 1.1 mmol) in methanol (8 mL, 16 mL, 32 mL) was added dropwise, the resulting reaction mixture was shaken and left overnight at room temperature with slow evaporation of the solvent. The next day, a crystalline precipitate was formed. The flask with the reaction mixture was tightly closed and kept at room temperature for another 5 days. Thereafter, the precipitate was filtered off and dried in vacuo to give ***(R,R/S,S)*-3**. Yield: 0.37 g (77.8%)—at initial concentration of thiourea **(±)-1** in a reaction mixture equal to 0.07 mol/L; 0.28 g (58.8%)—at initial concentration of thiourea **(±)-1** in a reaction mixture equal to 0.035 mol/L; 0.22 g (46.2%)—at initial concentration of thiourea **(±)-1** in a reaction mixture equal to 0.0175 mol/L. In all cases, the same product was isolated. M.p. 176–178 °C; IR (KBr): ν (cm⁻¹) 3184 (NH), 1548 (NCS), 1044, 1025 (C-O-P), 616 (P = S); ESI⁺-MS (DMF): *m/z* 773.2 [M + H]⁺, 359.1 [M + 2H-Ni-L]⁺; Elemental analysis calcd (%) for $C_{35}H_{49}N_5NiO_4P_2S_4$: C 49.30, H 5.79, N 8.21, Ni 6.88, P 7.26, S 15.04; found (%): C 49.48, H 5.60, N 7.97, Ni 6.65, P 7.04, S 15.16.

(R,R)-NiL₂·Py-type Complex ((R,R)-3): preparation method is the same as for ***(R,R/S,S)*-3** using ***(R)*-1** as initial thiourea (at initial concentration equal to 0.07 mol/L). Complex ***(R,R)*-3** was isolated by the same crystallization procedure as for ***(R,R/S,S)*-3**. Yield: 0.31 g (65.2%); m.p. 145–147 °C; $[\alpha]_D^{20}$ = +180 (c 0.5, C_6H_6); IR (KBr): ν (cm⁻¹) 3181 (NH), 1549 (NCS), 1047, 1026 (C-O-P), 615 (P = S); ESI⁺-MS (DMF): *m/z* 773.2 [M + H]⁺, 359.1 [M + 2H-Ni-L]⁺; Elemental analysis calcd (%) for $C_{35}H_{49}N_5NiO_4P_2S_4$: C 49.30, H 5.79, N 8.21, Ni 6.88, P 7.26, S 15.04; found (%): C 49.07, H 6.05, N 8.00, Ni 7.12, P 7.47, S 15.31.

(S,S)-NiL₂·Py-type Complex ((S,S)-3): preparation method is the same as for **(R,R/S,S)-3** using **(S)-1** as initial thiourea (at initial concentration equal to 0.07 mol/L). Complex **(S,S)-3** was isolated by the same crystallization procedure as for **(R,R/S,S)-3**. Yield: 0.3 g (63.1%); m.p. 146–147 °C; $[a]_D^{20}$ = −181 (c 0.5, C_6H_6); IR (KBr): ν (cm^{-1}) 3178 (NH), 1549 (NCS), 1046, 1026 (C-O-P), 614 (P = S); ESI⁺-MS (DMF): *m/z* 773.2 [M + H]⁺, 359.1 [M + 2H-Ni-L]⁺; Elemental analysis calcd (%) for $C_{35}H_{49}N_5NiO_4P_2S_4$: C 49.30, H 5.79, N 8.21, Ni 6.88, P 7.26, S 15.04; found (%): C 49.22, H 5.55, N 8.01, Ni 7.16, P 7.50, S 15.06.

2.2. X-ray Diffraction Study

Data sets for single crystals were collected on a Bruker AXS Kappa Apex diffractometer (Germany, Karlsruhe) with graphite-monochromated MoKα radiation (λ = 0.71073 Å). The structures were solved by direct methods using APEX3 [13] for data collection, SAINT [14] for data reduction, SHELXS [15] for structure solution, SHELXL [15] for structure refinement by full-matrix least-squares against F^2, and SADABS [16] for multi-scan absorption correction. Most of the crystals are of poor quality and exhibit positional disorder of the ethoxy-groups, which, for some crystals, was not possible to resolve, due to poor resolution. The corresponding fragments were refined isotropically. The poor quality of the crystals resulted in low accuracy of the geometrical parameters. Crystal **(R,R/S,S)-3** contains 5% of acetate ion coordinated to nickel and 95% of pyridine, the evidence for the presence of acetate ion is provided by the presence of two peaks in the vicinity of pyridine and the non-positive definite nitrogen atom of the pyridine moiety. The data collection and refinement parameters are given in Table 1. CCDC 1961489-1961494 contains the supplementary crystallographic data for this paper (Supplementary Materials).

Table 1. Crystallographic data and structure refinement details for compounds **1**–**3**.

Crystal	(R)-1	(±)-1	(R,R)-2	(meso)-2	(R,R)-3	(R,R/S,S)-3
Formula	$C_{15}H_{23}N_2O_2PS_2$	$C_{15}H_{23}N_2O_2PS_2$	$C_{30}H_{44}N_4NiO_4P_2S_4$	$C_{26}H_{40}N_4NiO_4P_2S_4$	$C_{35}H_{49}N_5NiO_4P_2S_4$	$C_{35}H_{49}N_5NiO_4P_2S_4$
CCDC number	1961490	1961489	1961491	1961492	1961494	1961493
Color	colorless	colorless	violet	violet	Green	green
Habitus	prizm	prizm	prizm	prizm	Prizm	prizm
Size (mm)	0.69 × 0.68 × 0.39	0.58 × 0.38 × 0.31	0.56 × 0.30 × 0.19	0.44 × 0.31 × 0.21	0.52 × 0.49 × 0.26	0.98 × 0.39 × 0.31
Formula weight	358.44	358.44	773.58	773.58	852.68	852.68
T (K)	150(2)	150(2)	100(2)	100(2)	100(2)	100(2)
Crystal system	monoclinic	triclinic	monoclinic	monoclinic	monoclinic	orthorhombic
Space group	$P2_1$	$P\text{-}1$	$P2_1$	$P2_1/c$	$P2_1$	$Pna2_1$
a (Å)	7.6020(10)	7.5459(9)	14.9354(6)	19.4428(14)	8.4467(11)	23.6547(15)
b (Å)	30.256(4)	8.1735(9)	7.5332(3)	12.3486(9)	20.960(3)	8.5751(5)
c (Å)	8.2300(11)	15.2017(16)	16.8715(7)	15.5240(11)	12.2036(17)	20.0009(13)
α (°)	90	92.278(6)	90	90	90	90
β (°)	108.423(5)	90.123(6)	112.192(2)	104.448(4)	107.971(6)	90
γ (°)	90	107.777(6)	90	90	90	90
V (Å³)	1795.9(4)	892.01(17)	1758.09(13)	3609.3(5)	2055.2(5)	4057.0(4)
Z	4	2	2	4	2	4
D Calcd (g m⁻³)	1.326	1.335	1.461	1.424	1.378	1.401
μ (mm⁻¹)	0.393	0.396	0.921	0.898	0.796	0.852
Reflection collected	55072	5016	30788	89973	35088	65953
Unique reflections	8927	5016	7769	8932	9905	9280
Reflections observed	8899	4517	6896	4555	8455	8111
θ min, θ max (°)	1.347, 28.346	1.341, 28.390	1.303, 27.236	1.081, 28.378	1.754, 28.375	1.722, 27.541
Goodness-of-fit (GOF) on F2	1.121	1.125	1.121	1.160	0.845	1.181
R1, wR2 (I ≥ 2σ(I))	0.0228, 0.0606	0.0392, 0.0952	0.0446, 0.1129	0.1345, 0.3012	0.0423, 0.1092	0.0547, 0.1540
R1, wR2 (all data)	0.0228, 0.0607	0.0454, 0.0970	0.0550, 0.1325	0.2452, 0.3519	0.0552, 0.1297	0.0643, 0.1620
Largest peak/hole (e Å⁻³)	0.296 / −0.248	0.389 / −0.266	1.321 / −0.943	0.749 / −1.132	0.539 / −0.859	1.553 / −0.575
Flack	0.007(6)		0.007(10)		0.013(17)	0.397(8)

3. Results and Discussion

The racemic and enantiopure thiophosphorylated thioureas **1** were synthesized by the addition reaction of the corresponding 1,2,3,4-tetrahydro-1-naphthylamine with O,O-diethyl thiophosphoryl isothiocyanate (Scheme 1). Square-planar complexes **2** and tetragonal pyramidal complexes **3** were obtained by the reactions of **1** with nickel(II) salts in the presence of potassium hydroxide (Scheme 2) and pyridine (Scheme 3), respectively. The syntheses of complexes **3** were carried out in the excess of pyridine, thus, one could expect pyridine to occupy both axial positions, however, no octahedral complexes were formed. To prove the exclusive formation of homochiral complexes from (±)-**1** in the presence of pyridine, the syntheses of (*R,R/S,S*)-**3** were carried out using different initial concentrations of precursors. The same products were obtained independent of concentrations.

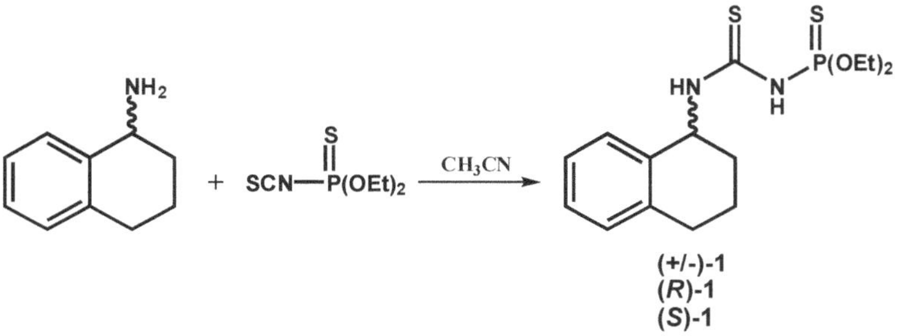

Scheme 1. The synthesis of N-thiophosphorylated thioureas **1**.

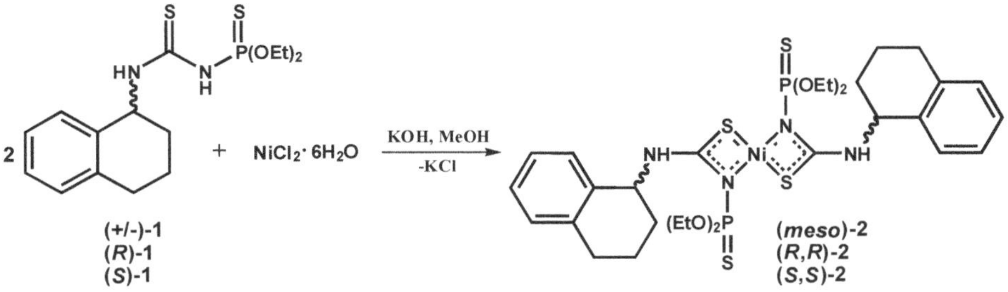

Scheme 2. The synthesis of square-planar complexes **2**.

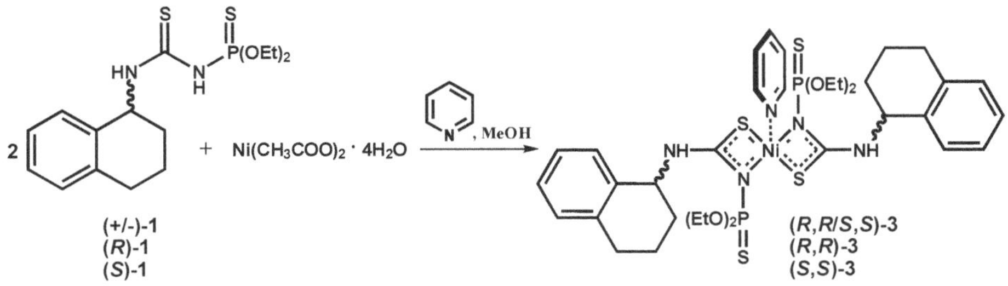

Scheme 3. The synthesis of tetragonal pyramidal complexes **3**.

Thiophosphorylated thioureas **1** are conformationally flexible compounds, which can adopt a variety of conformations very close on an energy scale [12]. Moreover, they exhibit nearly free internal rotation of the terminal ethoxy groups. In addition, they have several donor atoms able to coordinate metal ions, thereby they can display a variety of coordination modes in the complexes with transition metals, depending on the intramolecular interactions and the corresponding preferable conformations. As it was shown in a series of publications [17–25], the most abundant coordination mode of thiophosphorylated thioureas is 1,5-*S,S* metal coordination with the formation of the six-membered metal containing cyclic fragments, while the 1,3-*N,S* mode is rare [12,20,24,26,27].

Compound **1** in racemic and enantiopure crystals have similar molecular structures with two N–H bonds being *trans* to each other (Figure 1). Two sulfur atoms are also on opposite sides of the N–C–N–P fragment. Such a molecular structure ideally complies with the geometry requirements of metal complexes with 1,3-*N,S*-coordination. In a racemic crystal, the molecules form nearly planar centrosymmetric dimers via the N–H···S hydrogen bonding. This supramolecular synthon is very stable and is reproduced in enantiopure crystals through pseudocentrosymmetric arrangement of two crystallographically independent molecules (Figure 1).

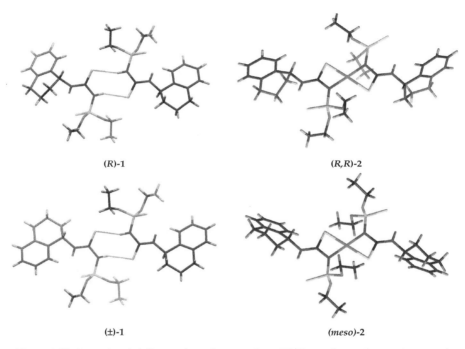

(*R*)-**1**   (*R,R*)-**2**

(±)-**1**   (*meso*)-**2**

Figure 1. Hydrogen bonded dimers of **1** and square-planar Ni(II) complexes **2** in enantiopure and racemic crystals.

Analyzing the geometry of hydrogen bonded dimers in the crystals of **1**, one can see an ideal preorganization of the ligands for 1,3-*N,S*-coordination of nickel (II) ions. Indeed, the X-ray single diffraction study shows the formation of 2:1 square-planar complexes with 1,3-*N,S*-coordination (Figure 1). Most important is that heterochiral centrosymmetric *meso*-complexes are formed from racemic ligands owing to the centrosymmetric Ni(II) coordination geometry. Interaction of (*R*)-**1** produced pseudocentrosymmetric homochiral complexes (*R,R*)-**2**. Worth mentioning is the transferability of the supramolecular geometry arrangement from dimeric hydrogen bonded synthon to the Ni(II) complex. One should note the equal distances between the donor atoms in dimers and in the complexes (Figure 1). The pseudocentrosymmetric planar arrangement of two molecules of (*R*)-**1** in homochiral dimers is quite distinct from the folded geometry of the dimers of the 1-phenylethyl-containing (*R*)-thiophosphorylated thioureas [12]. Interestingly, the crystallization of square-planar Ni(II) complexes of the latter yielded no single crystals that were suitable for X-ray diffraction analysis.

To break the symmetry of the Ni(II) coordination geometry, we have introduced an additional axial pyridine ligand (Scheme 3), which results in the exclusive formation of homochiral tetragonal pyramidal complexes from racemic ligands (Figure 2). Thus, the symmetry break via introduction of the axial ligand may be widely used to access homochiral complexes on a molecular level. Moreover, for racemic **3**, homochirality was achieved on a supramolecular level. In both crystals of racemic **3** and **(R,R)-3**, supramolecular homochiral chains are formed owing to key–lock steric interactions with pyridine ligands located in the cavity at the base of coordination pyramid of the neighbouring complex. The C–H⋯Ni contacts in crystals **3** are slightly longer (2.84–2.87 Å) than in 1-phenylethyl-containing complexes [12]. The molecules in a supramolecular chain are related via a translation operation. In racemic **3**, the supramolecular chains of opposite chirality interact via weak C–H⋯S and C–H⋯π interactions. These types of short contacts are also revealed in other crystals, depending on the conformations of the terminal ethoxy groups, the latter adopt *gauche* and *trans*-conformations (Table 2), with the torsional angles varying in wide limits.

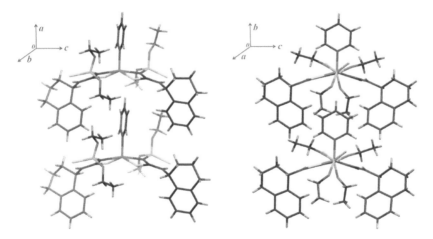

Figure 2. Molecular structure and the arrangement of molecules in homochiral supramolecu lar chains in the crystals of enantiopure **(R,R)-3** (**left**) and racemic **3** (**right**).

To conclude, new chiral thiophosphorylated thioureas were synthesized in racemic and enantiopure form. In racemic crystals, the molecules form nearly planar centrosymmetric dimers via the N−H⋯S hydrogen bonding. This supramolecular synthon is very stable and is reproduced in enantiopure crystals through pseudocentrosymmetric arrangement of two crystallographically independent molecules. The obtained thioureas exhibit the 1,3-*N,S*-coordination mode with Ni(II) and form 2:1 complexes. *Meso*-complexes are formed from racemic ligands with a centrosymmetric square-planar nickel coordination. Breaking the symmetry of nickel coordination geometry by the introduction of axial pyridine ligand results in homochiral complexes from racemic ligands. Molecular homochirality in tetragonal pyramidal complexes is further transferred to supramolecular homochiral arrangement via key–lock steric interactions. Thus, the presented approach allows to control the diastereoselectivity of complex formation without additional chiral auxiliaries. The key−lock homochiral supramolecular interactions show the perspective to obtain 1D-homochiral coordination polymers. Further studies on chirality control beyond the molecular level to achieve 3D supramolecular homochirality are in progress.

Table 2. Selected torsion angles (deg.) in the ligands and complexes.

Crystal	S=P–N–C	S=P–O–C P–O–C–C	S=P–O–C P–O–C–C	S=P–O–C P–O–C–C	S=P–O–C P–O–C–C	Pyridine Orientation
(*R*)-1 (mol A)	−56.0(2)	−33.2 (2) 109.2(2)	−48.1(2) 172.3(2)			
(*R*)-1 (mol B)	60.2(2)	47.0(2) −172.0(2)	33.8(2) −108.3(2)			
(±)-1	−55.2(5)	−49.8(5) −175.2(5)	−34.5(4) 111.3(5)			
(*R,R*)-2	36.6(5) 36.7(5)	38.6(4) −98.9(5)	49.0(4) −109.2(4)	−167.2(4) 106.2(6)	52.0(4) −170.4(4)	
(*meso*)-2 (mol A)	16.4(1) −16.4(1)	57.4(1) −163.1(8)	58.7(8) 165.4(2)	−57.4(1) 163.1(8)	−58.7(8) −165.4(2)	
(*meso*)-2 (mol B)	−19.9(1) 19.9(1)	−45.7(8) 163.3(2)	−56.1(1) −178.1(1)	56.1(1) 178.1(1)	45.7(8) −163.3(2)	
(*R,R*)-3	27.1(5) 1.1(4)	39.4(5) −116.5(5)	58.4(5) −164.2(5)	45.5(4) 87.6(6)	58.2(5) −178.5(4)	−24.3(4) −24.8(5)
(*R,R/S,S*)-3	−39.1(6) −44.5(5)	−52.9(6) −158.9(5)	−170.6(5) 165.2(5)	54.2(6) 172.6(5)	−178.6(5) −162.5(7)	−3.0(5) −11.7(6)

Supplementary Materials: The following are available online at http://www.mdpi.com/2073-4352/9/12/606/s1. CCDC 1961489-1961494 contains the supplementary crystallographic data for this paper. These data can be obtained free of charge from The Cambridge Crystallographic Data Centre.

Author Contributions: O.K. and K.M. prepared the manuscript with contributions of all other co-authors, R.Z., K.N., D.S., E.B. carried out synthetic work and crystal growth, Z.Y. and K.I. carried out X-ray single crystal diffraction studies, O.S. and V.A. are responsible for project administration.

Funding: This work was supported by the grant of the Russian Foundation for Basic Research, grant No 18-43-160018.

Acknowledgments: The authors gratefully acknowledge the Spectral-Analytical Center "CSF-SAC FRC KSC RAS" for providing necessary facilities to carry out this work. K. Ivshin is thankful to the Russian Government Program of Competitive Growth of Kazan Federal University.

Conflicts of Interest: The authors declare no conflict of interest.

References

1. Berthod, A. Chiral recognition mechanism. *Anal. Chem.* **2006**, *78*, 2093–2099. [CrossRef] [PubMed]
2. Bissantz, C.; Kuhn, B.; Stahl, M. A medicinal chemist's guide to molecular interactions. *J. Med. Chem.* **2010**, *53*, 5061–5084. [CrossRef] [PubMed]
3. Jędrzejewska, H.; Szumna, A. Making a right or left choice: Chiral self-sorting as a tool for the formation of discrete complex structures. *Chem. Rev.* **2017**, *117*, 4863–4899. [CrossRef] [PubMed]
4. Xing, P.; Zhao, Y. Controlling supramolecular chirality in multicomponent self-assembled systems. *Acc. Chem. Res.* **2018**, *51*, 2324–2334. [CrossRef]
5. Song, L.; Shemchuk, O.; Robeyns, K.; Braga, D.; Grepioni, F.; Leyssens, T. Ionic cocrystals of etiracetam and levetiracetam: The importance of chirality for ionic cocrystals. *Cryst. Growth Des.* **2019**, *19*, 2446–2454. [CrossRef]
6. Guijarro, A.; Yus, M. *The Origin of Chirality in the Molecules of Life: A Revision from Awareness to the Current Theories and Perspectives of this Unsolved Problem*, 1st ed.; Royal Society of Chemistry: Cambridge, UK, 2009; p. 160.
7. Tsang, M.Y.; Di Salvo, F.; Teixidor, F.; Viñas, C.; Planas, J.G.; Choquesillo-Lazarte, D.; Vanthuyne, N. Is molecular chirality connected to supramolecular chirality? The particular case of chiral 2-pyridyl alcohols. *Cryst. Growth Des.* **2015**, *15*, 935–945. [CrossRef]
8. Scriba, G.K.E. Recognition mechanisms of chiral selectors: An overview. In *Chiral Separations. Methods and Protocols*; Scriba, G.K.E., Ed.; Humana Press: Totowa, NY, USA, 2019; Volume 1985, pp. 1–33.

9. Sørensen, H.O.; Larsen, S. Hydrogen bonding in enantiomeric versus racemic mono-carboxylic acids; a case study of 2-phenoxypropionic acid. *Acta Crystallogr. B* **2003**, *59*, 132–140. [CrossRef]
10. Gavezzotti, A.; Lo Presti, L. Theoretical study of chiral carboxylic acids. Structural and energetic aspects of crystalline and liquid states. *Cryst. Growth Des.* **2015**, *15*, 3792–3803. [CrossRef]
11. Metlushka, K.E.; Sadkova, D.N.; Shaimardanova, L.N.; Nikitina, K.A.; Ivshin, K.A.; Islamov, D.R.; Kataeva, O.N.; Alfonsov, A.V.; Kataev, V.E.; Voloshina, A.D.; et al. First coordination polymers on the bases of chiral thiophosphorylated thioureas. *Inorg. Chem. Commun.* **2016**, *66*, 11–14. [CrossRef]
12. Kataeva, O.N.; Metlushka, K.E.; Yamaleeva, Z.R.; Ivshin, K.A.; Kiiamov, A.G.; Lodochnikova, O.A.; Nikitina, K.A.; Sadkova, D.N.; Punegova, L.N.; Voloshina, A.D.; et al. Co-ligand induced chiral recognition of *N*-thiophosphorylated thioureas in crystalline Ni(II) complexes. *Cryst. Growth Des.* **2019**, *19*, 4044–4056. [CrossRef]
13. Bruker. *APEX3 Crystallography Software Suite*; Bruker AXS Inc.: Madison, WI, USA, 2016.
14. Bruker. *SAINT. Crystallography Software Suite*; Bruker AXS Inc.: Madison, WI, USA, 2016.
15. Sheldrick, G.M. A Short history of SHELX. *Acta Crystallogr. A* **2008**, *64*, 112–122. [CrossRef]
16. Krause, L.; Herbst-Irmer, R.; Sheldrick, G.M.; Stalke, D. Comparison of silver and molybdenum microfocus X-ray sources for single-crystal structure determination. *J. Appl. Crystallogr.* **2015**, *48*, 3–10. [CrossRef]
17. Mitoraj, M.P.; Babashkina, M.G.; Robeyns, K.; Sagan, F.; Szczepanik, D.W.; Seredina, Y.V.; Garcia, Y.; Safin, D.A. Chameleon-like nature of anagostic interactions and its impact on metalloaromaticity in square-planar nickel complexes. *Organometallics* **2019**, *38*, 1973–1981. [CrossRef]
18. Safin, D.A.; Babashkina, M.G.; Robeyns, K.; Mitoraj, M.P.; Kubisiak, P.; Garcia, Y. Influence of the homopolar dihydrogen bonding C-H···H-C on coordination geometry: Experimental and theoretical studies. *Chem. Eur. J.* **2015**, *21*, 16679–16687. [CrossRef]
19. Safin, D.A.; Babashkina, M.G.; Bolte, M.; Mitoraj, M.P.; Klein, A. Complexing cation influences distortion of the ligand in the structure of [M{2-MeO(O)CC$_6$H$_4$NHC(S)NP(S)(OiPr)$_2$}$_2$] (M = Zn^{II}, Cd^{II}) complexes: A driving force for the intermolecular aggregation. *Dalton Trans.* **2015**, *44*, 14101–14109. [CrossRef]
20. Babashkina, M.G.; Safin, D.A.; Srebro, M.; Kubisiak, P.; Mitoraj, M.P.; Bolte, M.; Garcia, Y. Crucial influence of the intramolecular hydrogen bond on the coordination mode of RC(S)NHP(S)(OiPr)$_2$ in homoleptic complexes with Ni^{II}. *Eur. J. Inorg. Chem.* **2013**, *2013*, 545–555. [CrossRef]
21. Babashkina, M.G.; Safin, D.A.; Srebro, M.; Kubisiak, P.; Mitoraj, M.P.; Bolte, M.; Garcia, Y. Influence of CH$_2$Cl$_2$ for the structure stabilization of the Ni^{II} complex [Ni{6-MeO(O)CC$_6$H$_4$NHC(S)NP(S)(OiPr)$_2$-1,5-S,S'}$_2$]·CH$_2$Cl$_2$. *CrystEngComm* **2012**, *14*, 370–373. [CrossRef]
22. Babashkina, M.G.; Safin, D.A.; Robeyns, K.; Garcia, Y. Complexation properties of the crown ether-containing *N*-thiophosphorylated thiourea towards Ni^{II}. *Dalton Trans.* **2012**, *41*, 1451–1453. [CrossRef]
23. Safin, D.A.; Babashkina, M.G.; Bolte, M.; Klein, A. Versatile structures and photophysical properties of poly- and mononuclear Cu^{I} complexes with *N*-thiophosphorylated thioureas RNHC(S)NHP(S)(OiPr)$_2$ and phosphanes. *CrystEngComm* **2011**, *13*, 568–576. [CrossRef]
24. Babashkina, M.G.; Safin, D.A.; Bolte, M.; Srebro, M.; Mitoraj, M.; Uthe, A.; Klein, A.; Köckerling, M. Intramolecular hydrogen bonding controls 1,3-*N,S* vs. 1,5- *S,S'*-coordination in Ni^{II} complexes of *N*-thiophosphorylated thioureas RNHC(S)NHP(S)(OiPr)$_2$. *Dalton Trans.* **2011**, *40*, 3142–3153. [CrossRef]
25. Safin, D.A.; Babashkina, M.G.; Bolte, M.; Klein, A. Synthesis and characterization of [H$_2$NC(S)NP(S)(OiPr)$_2$]$^-$ complexes of Co(II), Ni(II), Zn(II) and Cd(II). *Inorg. Chim. Acta* **2011**, *365*, 32–37. [CrossRef]
26. Metlushka, K.E.; Sadkova, D.N.; Nikitina, K.A.; Lodochnikova, O.A.; Kataeva, O.N.; Alfonsov, V.A. Ni(II) complex of bisthiophosphorylated thiourea prepared from the Betti base. *Russ. J. Gen. Chem.* **2017**, *87*, 2130–2132. [CrossRef]
27. Safin, D.A.; Babashkina, M.G.; Mitoraj, M.P.; Kubisiak, P.; Robeyns, K.; Bolte, M.; Garcia, Y. An intermolecular pyrene excimer in the pyrene-labeled *N*-thiophosphorylated thiourea and its nickel(II) complex. *Inorg. Chem. Front.* **2016**, *3*, 1419–1431. [CrossRef]

Article

Identification of Structural Motifs Using Networks of Hydrogen-Bonded Base Interactions in RNA Crystallographic Structures

Hazrina Yusof Hamdani [1,*] and Mohd Firdaus-Raih [2,3,*]

[1] Advanced Medical and Dental Institute, Universiti Sains Malaysia, Bertam, Kepala Batas 13200, Pulau Pinang, Malaysia
[2] Faculty of Science and Technology, Universiti Kebangsaan Malaysia (UKM), Bangi 43600, Selangor, Malaysia
[3] Institute of Systems Biology, Universiti Kebangsaan Malaysia (UKM), Bangi 43600, Selangor, Malaysia
* Correspondence: hazrina@usm.my (H.Y.H.); firdaus@mfrlab.org (M.F.-R.)

Received: 11 September 2019; Accepted: 22 October 2019; Published: 24 October 2019

Abstract: RNA structural motifs can be identified using methods that analyze base–base interactions and the conformation of a structure's backbone; however, these approaches do not necessarily take into consideration the hydrogen bonds that connect the bases or the networks of inter-connected hydrogen-bonded bases that are found in RNA structures. Large clusters of RNA bases that are tightly inter-connected by a network of hydrogen bonds are expected to be stable and relatively rigid substructures. Such base arrangements could therefore be present as structural motifs in RNA structures, especially when there is a requirement for a highly stable support platform or substructure to ensure the correct folding and spatial maintenance of functional sites that partake in catalysis or binding interactions. In order to test this hypothesis, we conducted a search in available RNA crystallographic structures in the Protein Data Bank database using queries that searched for profiles of bases inter-connected by hydrogen bonds. This method of searching does not require to have prior knowledge of the arrangement being searched. Our search results identified two clusters of six bases that are inter-connected by a network of hydrogen bonds. These arrangements of base sextuples have never been previously reported, thus making this the first report that proposes them as novel RNA tertiary motifs.

Keywords: RNA structural motifs; base-base interactions; classification of base arrangement; RNA crystallographic structures

---

1. Introduction

Hydrogen bonds are crucial for stabilizing the complex structures of ribonucleic acids (RNA). Conformational changes in a particular RNA molecule can result from variations of the hydrogen bond interactions present in a structure. Clusters of unbroken networks of hydrogen-bonded base interactions have been reported previously in RNA structures [1–3]. In this work, a hydrogen-bonded base interaction network is defined as an unbroken connection of bases that are interacting with each other through at least one hydrogen bond. Even though such hydrogen-bonded base interaction networks have been reported before, there is a paucity of work discussing large inter-connected hydrogen-bonded clusters of bases in three-dimensional (3D) arrangements.

Various types of smaller base arrangements consisting of triples, quadruples, and quintuples that can form tertiary-level motifs (3D motifs) have been observed, reported, and archived [4–6]. However, to our knowledge, there has been no systematic study to identify larger arrangements composed of six base clusters and beyond, despite their possible importance as building-block modules of RNA structure. Identifying the existence of such 3D modules may lead to a more accurate design of functionally relevant synthetic RNA molecules in addition to improving the capacity to model RNA 3D structures from sequence information.

An RNA 3D motif can be defined as a tertiary arrangement of nucleotides, nucleosides, or bases that are repeatedly found in different locations either in the same RNA structure or in different RNA molecules. The annotation of RNA 3D motifs can be divided into three main approaches that consider: (i) the conformation of the RNA backbone [7,8]; (ii) the base–base interactions [1,3,9]; and (iii) the alignment of RNA 3D structures to detect similarities in folding and sub-folding [10,11]. Many of the known RNA 3D motifs discovered to date have an architectural role in RNA folding, functioning as stabilizers of RNA 3D structure or as sites of ligand binding or catalytic activity [12–14] More recently, RNA 3D motifs have been reported to play a role in miRNA biogenesis by acting as a guide for the Dicer-like 1 (DCL1) enzyme to perform cleavage of miRNA–miRNA* duplexes [15]. In this paper, we only focused on motifs that result from base–base interactions.

Expert visual examination of RNA structures has been a crucial aspect in the discovery of many currently known 3D base motifs [4,16]. RNA 3D motifs that have been discovered and annotated are available in various databases such as NCIR [4], RNA 3D Motif Atlas [9], and INTERRNA [6]. Several computer programs are available for identifying motifs in the available dataset of RNA crystallographic structures [10,17–19], and these programs allow for an automated search capacity that overcomes the limitations of manual visual curation such as the approach used for the NCIR database.

However, many of these computer programs require prior knowledge of the motifs to be provided as search queries, thus making them useful for structural annotation purposes but of limited utility for the discovery of novel motifs. The computer program COGNAC (COnnection tables Graphs for Nucleic Acids) was reported to be able to search for unbroken networks of hydrogen-bonded base interactions in RNA crystallographic structures that are available in the Protein Data Bank PDB [1]. A COGNAC search relies only on the user defining the connectivity of the bases by hydrogen bonds without the need for specific prior knowledge of how the bases are arranged in 3D space. Due to this capability, the COGNAC program may retrieve arrangements that are potentially novel motifs.

COGNAC annotations are independent of base sequence and any specific spatial arrangement definitions, thus making it possible to retrieve similar base components that, when visually examined, are in fact different in terms of spatial base arrangements. For example, the annotations for a query composed of GGGG, that was annotated as a planar base quadruple in the spinach RNA aptamer (G72.G29.G25.G68) (PDB ID: 4TS0), differed significantly from the GGGG base quadruple identified in the *Kluyveromyces lactis* 80S ribosomal structure (G1433.G1278.G1273.G1277) (PDB ID: 4V92) that was retrieved by the same search (Figure 1) [20,21]. While it is clear that the COGNAC program could potentially identify novel motifs, a secondary classification technique would be needed to sift through the hydrogen bond-connected base clusters that it retrieved.

ID	Sequence	Bases in Interaction Chain_BaseResidueBaseNumber	Structure Description	PDBID
QUAD1_67767	GGGG	Y_G72: X_G29: X_G25: Y_G68	Crystal structure of the Spinach RNA aptamer in complex with DFHBI barium ions	4ts0
QUAD1_70860	GGGG	A_G1433: A_G1278: A_G1273: A_G1277	Kluyveromyces lactis 80S ribosome in complex with CrPV-IRES	4V92

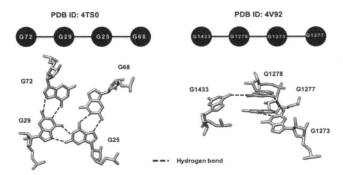

Figure 1. An example of a GGGG arrangement found in the structure of a spinach RNA aptamer (PDB ID: 4TS0) and *Kluyveromyces lactis* 80S ribosome (PDB ID: 4V92) that have different tertiary base arrangements, despite being retrieved using the same graph representation query of the COGNAC computer program.

In this paper, we report the identification of 3D motifs in RNA structures by analyzing the data of clusters of base interactions involving six bases that are interconnected by hydrogen bonds. Our results revealed six base clusters that are repeated in different RNA molecules or can be found at different locations in the same molecules, which makes them possible tertiary motifs.

2. Materials and Methods

2.1. Dataset

In this study, 2158 structure coordinates containing RNA chains that were solved by X-ray diffraction of crystals were downloaded from the Protein Data Bank (PDB) [22]. This dataset includes coordinate files that contain RNA chains in the presence of other macromolecules or ligands. The downloaded structures represent a diverse repertoire of the available RNA molecules such as rRNA, ribozymes, riboswitches, mRNA, and tRNA. The resolution cutoff of the structures is ≤4 Å.

2.2. COGNAC Searches

The hydrogen bonding data for the 2158 structures were then generated by HBPRED, a program that was reported by Firdaus-Raih et al. [1]. The HBPRED program is based on the hydrogen bonding parameters used in the HBPLUS program but with specific modifications for use with RNA bases [23]. The hydrogen bonding information for the bases in all 2158 structures was then searched for specific arrangements of six base clusters (sextuples), where each base in a cluster is connected to another base by at least one hydrogen bond, using the COnnection tables Graphs for Nucleic ACids (COGNAC) computer program. The graph representations for the six possible arrangements of a base in a sextuple, each base being connected to at least one other base via at least one hydrogen bond (Table 1), were previously described by Firdaus-Raih et al. [1].

Table 1. Graph representations of six possible base connectivity patterns of a sextuple, where a base is connected to at least one other base by at least one hydrogen bond, and the number of occurrences found in the search dataset for each base sextuple type.

Pattern	Number of Annotations	Number of Unique Arrangements	Pattern	Number of Annotations	Number of Unique Arrangements
Type 1	10168	3572	Type 2	5448	2102
Type 3	4240	1862	Type 4	525	299
Type 5	8	6	Type 6	614	311

2.3. Sextuple Sub-Classification

The results generated from the COGNAC searches for each type of sextuple were deposited into the Interactions in RNA Structures Database (InterRNA) and can be accessed at http://mfrlab.org/interrna/. Each database entry is retrievable using an identification code referred to as the INTERRNA ID [6]. The flowchart for the data analysis and filtering process to sub-classify each type of sextuple is provided in Figure 2a.

The hydrogen bonding data for the COGNAC annotations were then parsed and extracted into matrices using a Perl program. The matrix for each hydrogen bond-connected base cluster contains information on the hydrogen bond donor base, the hydrogen bond donor atom, the hydrogen bond acceptor base, and the hydrogen bond acceptor atom. The matrices were compared to identify a unique set of hydrogen-bonded base networks. The extracted data for the hydrogen-bonded base networks served as the input for the next phase to sub-classify each sextuple.

Next, the search queries and all the hydrogen-bonded base clusters were compared, and each sextuple was further sub-classified using the PHP programming language. This phase of the data processing involved the extraction of unique base clusters from the COGNAC searches; the pseudocode for the process is provided in Figure 2b. Manual visual examinations of the structures were carried out using the UCSF Chimera molecular graphics suite [24]. These analyses included assessments of the different arrangements found superposed against each other using least-squares superpositions. Follow-up multiple-sequence alignments were carried out to detect the conservation of the observed sextuples at the sequence level.

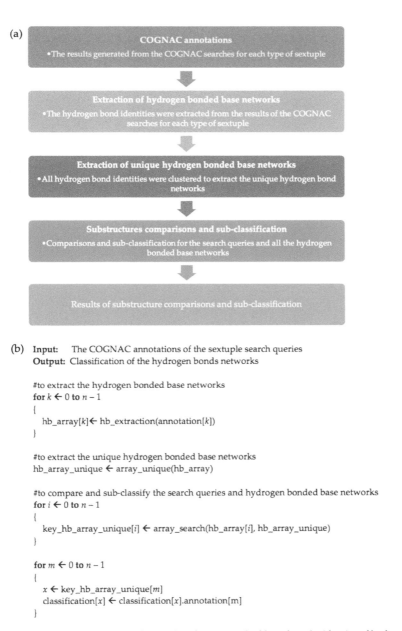

Figure 2. (**a**) Flowchart of the sextuple sub-classification method based on the identity of hydrogen bonding; (**b**) pseudo-code of the classification technique based on the identity of hydrogen bonding.

3. Results and Discussion

3.1. Filtering of COGNAC Searches

The methodology that employs the COGNAC computer program used in this study has been proven to be able to identify hydrogen bond-connected bases from pairings to interactions involving six bases [1,6]. Although the COGNAC approach was reported to be able to extract novel motifs without the need for prior knowledge of the tertiary base arrangements involved, it has the distinct limitation of being able to only search for arrangements in which the bases are connected by hydrogen bonds. That limitation, however, suited the ambit of this work in trying to find clusters of bases that are interconnected by hydrogen bonds.

This work further limited the scope of analysis and discussion to only sextuple patterns (Table 1) because smaller hydrogen bond-connected base arrangements such as triples, quadruples, and quintuples have been reported and are also likely to be constituents of a larger sextuple interaction. For example, the quadruple interaction (U438.A496.A498.G404) is a component of the sextuple interaction (U438.A496.A498.G404.A499.U439) annotated in the 16S rRNA of *Escherichia coli* (PDB ID: 5J7L) [25].

The COGNAC searches returned a total of are 3572 unique arrangements that were extracted from 10,168 annotations for the Type 1 sextuple pattern, 2102 pattern arrangements that were classified from 5448 annotations for the Type 2 sextuple pattern, and 1862 pattern arrangements that were classified from 4240 annotations for the Type 3 sextuple pattern. Furthermore, we obtained 299 pattern arrangements that were classified from 525 annotations for the Type 4 sextuple pattern, 6 pattern arrangements that were classified from 8 annotations for the Type 5 sextuple pattern, and 311 pattern arrangements that were classified from 614 annotations for the Type 6 sextuple pattern (Table 1).

Once the COGNAC search results for each sextuple type were sub-classified, the arrangements were analyzed to identify potentially novel structural motifs. In order to be considered as a potential motif, we required that the candidate arrangement fulfilled at least one of two criteria. The first criterion to be considered a motif was that the arrangement was repeatedly present in different RNA molecules, while the second criterion, especially if the first criterion was not met, was that an arrangement could be found in different locations of the same structure.

3.2. Hydrogen Bond-Connected Six-Base Interactions as Novel Structural Motifs

The results of the COGNAC searches were manually curated and were followed up by extensive visual examination to confirm the fitness of the patterns retrieved to our search criteria. Through this process, we were able to identify two sextuple base clusters that we propose as novel structural motifs, one meeting the first criterion, and the other fitting the second criterion [24].

3.2.1. A Base Sextuple Annotated in Different RNA Structures

One potentially novel motif that was uncovered by our visual examination process of the filtered COGNAC results is a GCA(A/U)(U/A)A Type 1 sextuple (Figure 3). The hydrogen bond donor–acceptor combinations (Figure 3a) clearly fitted that of a Type 1 sextuple pattern (Figure 3c). Our searches and analysis found this arrangement in four different RNA structures, i.e., 16S rRNA, 5S rRNA, a preQ1 riboswitch, and a S-adenosyl-(L)-homocysteine (SAH) riboswitch (Figure 3).

In this particular case, the base components at the R1, R2, R3, and R6 positions are conserved, with R1–R3 superposing very well onto each other (Figure 3c). However, in three structures, the R5 position is a uracil, while the R4 is an adenine, but in another structure the R4 is a uracil, while the R5 is an adenine. Despite this variation, it is clear that the 3D space taken up by both R4 and R5 are conserved due to the use of the same base components, despite them being on opposing sides of the pairing. This suggests that such a sextuple can be a structural module that can occur within specific spatial constraints and that variation in the sequence can exist, although the structures seem to be unrelated at sequence level (purine to pyrimidine). This is actually a practical mechanism for increasing sequence diversity that will at the same time maintain structural conservation, because the variations occur as a pair (AU to UA) that retains the interaction space of the pairing that was replaced.

Further scrutiny and visual examination of each occurrence of this GCA(A/U)(U/A)A Type 1 sextuple revealed that these motifs are found in parts of their respective RNA molecules that do not seem to share the same roles in different molecules. The GCA(A/U)(U/A)A Type 1 sextuple 3D motif, found in the 16S rRNA (G113.C314.A51.U114.A313.A116), is situated in a location that does not appear to be directly involved in peptidyl-transferase activity. However, it is likely that this six-base cluster contributes to the structural stability of the ribosomal small subunit and, therefore, has an indirect role in protein synthesis. By visual examination of the cluster, we noted that it connects helix 5 and helix 7 to the body of the 16S rRNA structure, where helix 7 is one of three helices (helix 7, helix 44, and helix 16/17) that serve as the 'structural pole' of the 16S rRNA molecule [26]. We further observed that this potential 3D motif is located near and adjacent to bases that are involved in the interactions of the ribosomal proteins S12 and S16 [26,27].

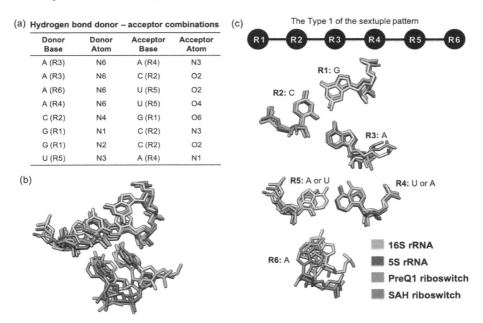

Figure 3. An example of a GCA(A/U)(U/A)A 3D arrangement that was annotated in four different RNA crystal structures. (**a**) Table of hydrogen bond donors and acceptors that define the hydrogen-bonded base network; (**b**) superposition of four similar base arrangements showing how they can be viewed in the four different RNA crystal structures; (**c**) graphical representation of the hydrogen bond network for the type 1 sextuple pattern (top) and corresponding base-by-base view of the superposed bases in the four different RNA crystal structures (bottom).

In the 5S rRNA structure, the G53.C29.A56.A54.U28.A57 cluster is located in loop B of domain I, which is between loop C of helix II and helix III that interact with Domain V of the 23S rRNA, with the presence of ribosomal proteins L5 and L18 (Figure 4b). The L5 ribosomal protein also interacts with the ribosomal protein subunit S13, and subsequently, the C terminal region of the ribosomal protein subunit S13 is located between the anticodon arms of the A- and P-tRNA sites, which functions as the nucleic acid decoding center; these sites are collectively known as the protuberance center [28]. This G53.C29.A56.A54.U28.A57 cluster does not appear to be directly involved in the protuberance center but more likely serves to reinforce the stability of the 5S rRNA (Domain 1).

In the preQ1 riboswitch structure, the G2.C21.A25.A3.U20.A26 cluster is located on the S1 stem and the L3 loop (Figure 4c). This particular Pre Q1 riboswitch is a 34-nucleotide H-type pseudoknot structure from *Bacillus subtilis* that serves as the preQ1 ligand recognition site [29]. In general, the H-type pseudoknot is folded via the interior of the hairpin loop, forming an intra-molecular interaction with the base of the exterior stem and resulting in a pseudoknot with two stems and two loops [30].

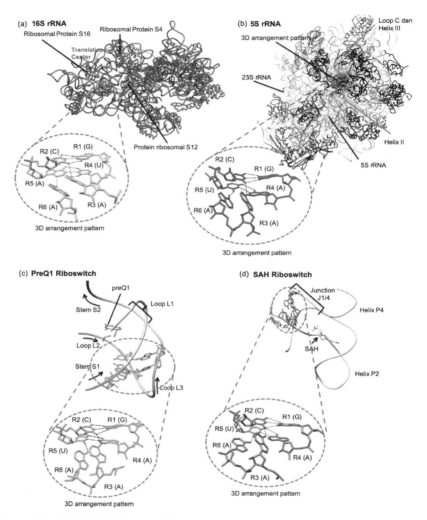

Figure 4. The GCA(A/U)(U/A)A motif (**a**) in the 16S rRNA crystal structure represented as orange spheres and magnified to show the base arrangements using orange stick representations (PDB ID: 1IBM) [27]; (**b**) in the 5S rRNA crystal structure represented as blue spheres and magnified to show the base arrangements using blue stick representations (PDB ID: 1JJ2) [31]; (**c**) in the preQ1 riboswitch crystal structure represented as purple sticks with the arrangement extracted and magnified (PDB ID: 3K1V) [29]; (**d**) in the SAH riboswitch crystal structure represented as green sticks with the arrangement extracted and magnified (PDB ID: 3NPQ) [32].

However, the H-type pseudoknot found in the 3K1V structure has two stems and three loops [29]. According to Aalberts and Hodas, the L3 loop where the G2.C21.A25.A3.U20.A26 cluster is annotated is a frequent location of many tertiary interactions [33]. There is also a quintuple base interaction, U6.A29.C18.G5.A28 (InterRNA ID: QUIN1_10653), very closely located at L1, end of S1, and end of L3, which has been identified as the ligand recognition site [29]. It is therefore highly possible that the G2.C21.A25.A3.U20.A26 cluster contributes to the stability of the ligand recognition site because it is located under the recognition site quintuple.

The fourth GCA(A/U)(U/A)A Type 1 sextuple can be found in the structure of a SAH riboswitch (G5.C34.A43.A6.U33.A44) (Figure 4d). The SAH riboswitches bind to the S-Adenosyl-l-homocysteine (SAH) molecule and can be found in many bacterial species [32]. The SAH riboswitch crystal structure consists of three helices, named P1, P2, and P3 [32]. The six-base cluster annotated in the SAH riboswitch is located at the P1 helix and the J1/4 junction [32], and our observations show that the sextuple does not interact with the SAH molecule.

However, two adenine bases (A43.A44) at the J1/4 junction form hydrogen bonds with the other four in the P1 helix, thus making the structure appear to be a type LL pseudoknot [32]. As with the previous observations regarding this sextuple, we also believe that it serves to increase the rigidity of the overall structure and is not directly associated with SAH binding that defines this riboswitch [32].

The structures in which the GCA(A/U)(U/A)A Type 1 sextuple could be found are not functionally related. Although the six nucleotides involved in this motif are closely situated in three-dimensional space, they span different sequence lengths and thus are not comparable using multiple-sequence alignments.

3.2.2. A Base Sextuple Annotated at Different Locations in the Same Structure

In addition to the six-base cluster that we found in four different RNA molecules, we were also able to identify a UAAGAC Type 2 sextuple that was annotated in three different locations in the same large ribosomal subunit structures, thus fitting our second criterion of base arrangements that could be potentially classified as tertiary motifs (Figure 5). This six-base cluster has a hydrogen bonded base connectivity pattern (Figure 6a) that fitted the Type 2 sextuple arrangement (Figure 6c).

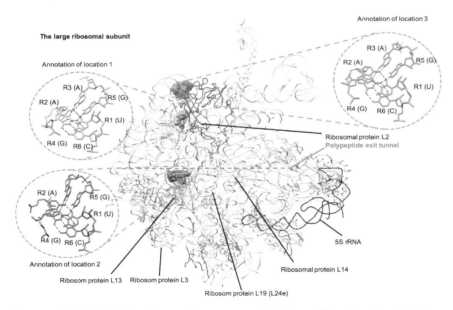

Figure 5. The UAAGAC base clusters that were annotated in three different locations in the large ribosomal subunit crystal structures are presented as colored spheres (PDB ID: 4LT8) [34].

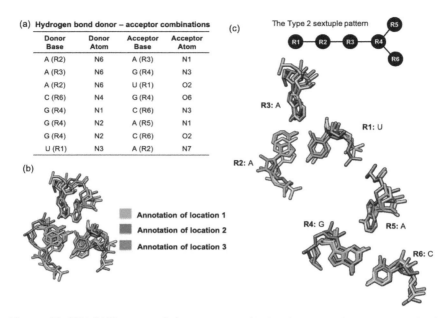

Figure 6. The UAAGAC base sextuple that was annotated in three locations in the same RNA molecule, a large ribosomal subunit structure (PDB ID: 4LT8). (**a**) Table of the hydrogen bond donors and acceptors that define the hydrogen-bonded base network; (**b**) superposition of three similar base arrangements as they can be viewed in the three different locations in same ribosomal subunit structure; (**c**) graphical representation of the hydrogen bond network for the Type 2 sextuple pattern and the corresponding base-by-base view of the superposed bases in the three different locations of the ribosomal subunit structure.

The first location of this UAAGAC sextuple is at helix 52 and helix 56 of Domain III in the large ribosomal subunit [35,36]. We also noted that this first site is located near the L2 ribosomal protein, which is one of the ribosomal proteins nearest to the peptidyl transferase center [37,38].

The second UAAGAC sextuple is located at Domain IV, helix 61, and helix 62 [35,36]. This second location is situated near two ribosomal proteins, L3 and L14 (Figure 5). The L3 ribosomal protein is one of the closest to the peptidyl transferase center [37,38]. In addition to the ribosomal protein L3, the ribosomal protein L14 is also involved indirectly in the binding site [39].

The third location for this UAAGAC sextuple, based on the large ribosomal subunit reference structures of *E. coli* and *Thermus thermophilus*, is at Domain III, helix 55, and helix 58 [35,36]. This UAAGAC sextuple is close to the ribosomal protein L2 [38]. However, we noted that this site could only be annotated in the large ribosomal subunit crystal structures of bacteria and was not found in the available examples for archaea.

Multiple-sequence alignments of the large ribosomal subunit of *Haloarcula marismortui*, *E. coli*, and *T. thermophiles* confirmed the results of the structural analysis that the third location for this UAAGAC Type 2 sextuple does not appear to be present in the archaea example (Figure 7). The third location appears to be in a more variable region in the structure of the large ribosomal subunit of *H. marismortui*. As with the previous GCA(A/U)(U/A)A Type 1 sextuple, the UAAGAC sextuple also appears to play a role in structural stabilization and does not appear to be directly involved in catalysis or as a binding site.

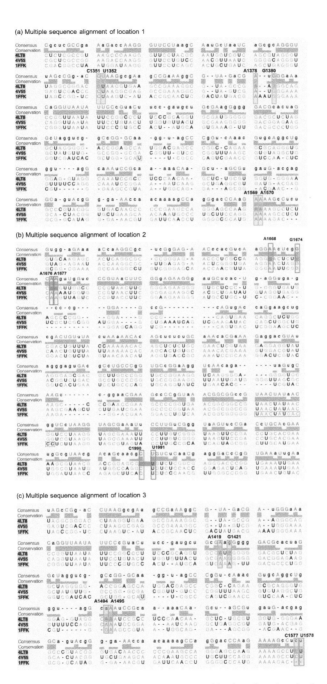

Figure 7. The UAAGAC base sextuples that were annotated in three locations in the sequences of the same RNA molecule, the large ribosomal subunit structures of *H. marismortui* (PDB ID: 1FFK), *T. thermophilus* (PDB ID: 4LT8), and *E. coli* (PDB ID: 4V55). (**a**) Multiple-sequence alignment of location 1 (partial sequences shown); (**b**) multiple-sequence alignment of location 2 (partial sequences shown) (**c**) multiple-sequence alignment of location 3 (partial sequences shown).

3.3. Presence of Known Motifs within the Sextuples

The sextuples that we report here are actually composed of elements that are already well known as tertiary motifs. For example, both sextuples contain A-minor motifs [16]. An AGC type I A-minor motif can be found in the GCA(A/U)(U/A)A Type 1 sextuple, while an AGC type II A-minor motif is a component of the UAAGAC Type 2 sextuple. Furthermore, the presence of other structural motifs, such as kink turns, can be seen and are expected to also be a feature of such larger motifs.

It is likely that other similarly large novel motifs discovered in the future will be composed of smaller known motifs. In this work, we demonstrate that, although the sextuples have known motifs as their constituent parts, mere visual examination was inadequate at identifying that they in fact partake in a larger conserved tertiary arrangement.

It is also worth noting that the COGNAC searches aimed at identifying novel motifs, as in this work, were intended for detecting tertiary motifs with the prerequisite of the component bases being connected by at least one hydrogen bond. It is therefore possible that similar arrangements may exist that were not retrieved by our search because they did not satisfy the hydrogen bonding criterion set. However, other programs that can be used to identify such motifs are already available and can be integrated in the process here reported, should they be required.

4. Conclusions

A cluster of six bases that are interconnected by a network of hydrogen bonds is expected to be a highly stable sub-structure. Due to this, it is unsurprising that our examination of the structures in which the GCA(A/U)(U/A)A Type 1 and the UAAGAC Type 2 sextuples can be found revealed a structural stabilization role that contributes to the correct folding and tertiary space maintenance of the associated functional sites. To our knowledge, this is the first such report that identifies these six bases as an associated cluster. Therefore, we propose that these base sextuples be classified as novel RNA tertiary motifs that may even have a wider role as RNA structural modules.

Author Contributions: Conceptualization, M.F.-R. Methodology and Software, M.F.-R. and H.Y.H.; Validation, H.Y.H.; Formal Analysis, H.Y.H.; Investigation, H.Y.H.; Resources, H.Y.H.; Writing—Original Draft Preparation, H.Y.H.; Writing—Reviewing & Editing, M.F.-R.; Visualization, H.Y.H.; Supervision, M.F.-R.; Project Administration, M.F.-R.; Funding Acquisition, H.Y.H. and M.F.-R.

Funding: This research was funded by the Universiti Sains Malaysia grant 304/CIPPT/6315258 to H.Y.H. and the Universiti Kebangsaan Malaysia grants DIP-2017-013 and DIP-2019-016 to M.F.-R.

Acknowledgments: We thank the Malaysia Genome Institute, Universiti Kebangsaan Malaysia and Universiti Sains Malaysia for computing resources. Molecular graphics and analyses were performed using UCSF Chimera, developed by the Resource for Biocomputing, Visualization, and Informatics at the University of California, San Francisco, with support from NIH P41-GM103311.

Conflicts of Interest: The authors declare no conflict of interest.

References

1. Firdaus-Raih, M.; Hamdani, H.Y.; Nadzirin, N.; Ramlan, E.I.; Willett, P.; Artymiuk, P.J. COGNAC: A Web Server for Searching and Annotating Hydrogen-Bonded Base Interactions in RNA Three-Dimensional Structures. *Nucleic Acids Res.* **2014**, *42*, W382–W388. [CrossRef]
2. Hamdani, H.Y.; Artymiuk, P.J.; Firdaus-Raih, M. A Computational Approach for the Annotation of Hydrogen-Bonded Base Interactions in Crystallographic Structures of the Ribozymes. *AIP Conf. Proc.* **2015**, *1678*. [CrossRef]
3. Bhattacharya, S.; Jhunjhunwala, A.; Halder, A.; Bhattacharyya, D.; Mitra, A. Going beyond Base-Pairs: Topology-Based Characterization of Base-Multiplets in RNA. *RNA* **2019**, *25*, 573–589. [CrossRef]
4. Nagaswamy, U. NCIR: A Database of Non-Canonical Interactions in Known RNA Structures. *Nucleic Acids Res.* **2002**, *30*, 395–397. [CrossRef]
5. Firdaus-Raih, M.; Harrison, A.M.; Willett, P.; Artymiuk, P.J. Novel Base Triples in RNA Structures Revealed by Graph Theoretical Searching Methods. *BMC Bioinform.* **2011**, *12* (Suppl. 13). [CrossRef]

6. Appasamy, S.D.; Hamdani, H.Y.; Ramlan, E.I.; Firdaus-Raih, M. InterRNA: A Database of Base Interactions in RNA Structures. *Nucleic Acids Res.* **2016**, *44*, D266–D271. [CrossRef] [PubMed]
7. Ren, H.; Shen, Y.; Zhang, L. The λ-Turn: A New Structural Motif in Ribosomal RNA. In *Lecture Notes in Computer Science (Including Subseries Lecture Notes in Artificial Intelligence and Lecture Notes in Bioinformatics)*; Springer: Berlin/Heidelberg, Germany, 2015; Volume 9226, pp. 456–466. [CrossRef]
8. Shen, Y.; Zhang, L. The Hasp Motif: A New Type of RNA Tertiary Interactions. In *Lecture Notes in Computer Science (Including Subseries Lecture Notes in Artificial Intelligence and Lecture Notes in Bioinformatics)*; Springer: Berlin/Heidelberg, Germany, 2017; Volume 10362 LNCS, pp. 441–453. [CrossRef]
9. Parlea, L.G.; Sweeney, B.A.; Hosseini-Asanjan, M.; Zirbel, C.L.; Leontis, N.B. The RNA 3D Motif Atlas: Computational Methods for Extraction, Organization and Evaluation of RNA Motifs. *Methods* **2016**, *103*, 99–119. [CrossRef] [PubMed]
10. Zahran, M.; Sevim Bayrak, C.; Elmetwaly, S.; Schlick, T. RAG-3D: A Search Tool for RNA 3D Substructures. *Nucleic Acids Res.* **2015**, *43*, 9474–9488. [CrossRef] [PubMed]
11. Piątkowski, P.; Jabłónska, J.; Zyła, A.; Niedziałek, D.; Matelska, D.; Jankowska, E.; Waleń, T.; Dawson, W.K.; Bujnicki, J.M. SupeRNAlign: A New Tool for Flexible Superposition of Homologous RNA Structures and Inference of Accurate Structure-Based Sequence Alignments. *Nucleic Acids Res.* **2017**, *45*, e150. [CrossRef]
12. Hendrix, D.K.; Brenner, S.E.; Holbrook, S.R. RNA Structural Motifs: Building Blocks of a Modular Biomolecule. *Q. Rev. Biophys.* **2005**, *38*, 221–243. [CrossRef]
13. Schroeder, K.T.; Mcphee, S.A.; Ouellet, J.; Lilley, D.M.J. A Structural Database for K-Turn Motifs in RNA. *RNA* **2010**, *16*, 1463–1468. [CrossRef] [PubMed]
14. Lilley, D.M.J. The K-Turn Motif in Riboswitches and Other RNA Species. *Biochim. Biophys. Acta Gene Regul. Mech.* **2014**, *1839*, 995–1004. [CrossRef] [PubMed]
15. Miskiewicz, J.; Szachniuk, M. Discovering Structural Motifs in MiRNA Precursors from the Viridiplantae Kingdom. *Molecules* **2018**, *23*, 1367. [CrossRef] [PubMed]
16. Nissen, P.; Ippolito, J.A.; Ban, N.; Moore, P.B.; Steitz, T.A. RNA Tertiary Interactions in the Large Ribosomal Subunit: The A-Minor Motif. *Proc. Natl. Acad. Sci. USA* **2001**, *98*, 4899–4903. [CrossRef]
17. Lai, C.E.; Tsai, M.Y.; Liu, Y.C.; Wang, C.W.; Chen, K.T.; Lu, C.L. FASTR3D: A Fast and Accurate Search Tool for Similar RNA 3D Structures. *Nucleic Acids Res.* **2009**, *37* (Suppl. 2), W287–W295. [CrossRef]
18. Hamdani, H.Y.; Appasamy, S.D.; Willett, P.; Artymiuk, P.J.; Firdaus-Raih, M. NASSAM: A Server to Search for and Annotate Tertiary Interactions and Motifs in Three-Dimensional Structures of Complex RNA Molecules. *Nucleic Acids Res.* **2012**, *40*, W35–W41. [CrossRef]
19. Yen, C.Y.; Lin, J.C.; Chen, K.T.; Lu, C.L. R3D-BLAST2: An Improved Search Tool for Similar RNA 3D Substructures. *BMC Bioinform.* **2017**, *18*, 574. [CrossRef]
20. Warner, K.D.; Chen, M.C.; Song, W.; Strack, R.L.; Thorn, A.; Jaffrey, S.R.; Ferré-D'Amaré, A.R. Structural Basis for Activity of Highly Efficient RNA Mimics of Green Fluorescent Protein. *Nat. Struct. Mol. Biol.* **2014**, *21*, 658–663. [CrossRef]
21. Fernández, I.S.; Bai, X.C.; Murshudov, G.; Scheres, S.H.W.; Ramakrishnan, V. Initiation of Translation by Cricket Paralysis Virus IRES Requires Its Translocation in the Ribosome. *Cell* **2014**, *157*, 823–831. [CrossRef]
22. Burley, S.K.; Berman, H.M.; Bhikadiya, C.; Bi, C.; Chen, L.; Di Costanzo, L.; Christie, C.; Dalenberg, K.; Duarte, J.M.; Dutta, S.; et al. RCSB Protein Data Bank: Biological Macromolecular Structures Enabling Research and Education in Fundamental Biology, Biomedicine, Biotechnology and Energy. *Nucleic Acids Res.* **2019**, *47*, D464–D474. [CrossRef]
23. McDonald, I.K.; Thornton, J.M. Satisfying Hydrogen Bonding Potential in Proteins. *J. Mol. Biol.* **1994**, *238*, 777–793. [CrossRef] [PubMed]
24. Pettersen, E.F.; Goddard, T.D.; Huang, C.C.; Couch, G.S.; Greenblatt, D.M.; Meng, E.C.; Ferrin, T.E. UCSF Chimera—A Visualization System for Exploratory Research and Analysis. *J. Comput. Chem.* **2004**, *25*, 1605–1612. [CrossRef] [PubMed]
25. Cocozaki, A.I.; Altman, R.B.; Huang, J.; Buurman, E.T.; Kazmirski, S.L.; Doig, P.; Prince, D.B.; Blanchard, S.C.; Cate, J.H.D.; Ferguson, A.D. Resistance Mutations Generate Divergent Antibiotic Susceptibility Profiles against Translation Inhibitors. *Proc. Natl. Acad. Sci. USA* **2016**, *113*, 8188–8193. [CrossRef] [PubMed]
26. Schluenzen, F.; Tocilj, A.; Zarivach, R.; Harms, J.; Gluehmann, M.; Janell, D.; Bashan, A.; Bartels, H.; Agmon, I.; Franceschi, F.; et al. Structure of Functionally Activated Small Ribosomal Subunit at 3.3 Å Resolution. *Cell* **2000**, *102*, 615–623. [CrossRef]

27. Ogle, J.M.; Brodersen, D.E.; Clemons, W.M., Jr.; Tarry, M.J.; Carter, A.P.; Ramakrishnan, V. Recognition of Cognate Transfer RNA by the 30S Ribosomal Subunit. *Science* **2001**, *292*, 897–902. [CrossRef]

28. Dinman, J.D. 5S RRNA: Structure and Function from Head to Toe. *Int. J. Biomed. Sci.* **2005**, *1*, 2–7. [PubMed]

29. Klein, D.J.; Edwards, T.E.; Ferré-D'Amaré, A.R. Cocrystal Structure of a Class I PreQ1 Riboswitch Reveals a Pseudoknot Recognizing an Essential Hypermodified Nucleobase. *Nat. Struct. Mol. Biol.* **2009**, *16*, 343–344. [CrossRef]

30. Staple, D.W.; Butcher, S.E. Pseudoknots: RNA Structures with Diverse Functions. *PLoS Biol.* **2005**, *3*, e213. [CrossRef]

31. Klein, D.J.; Schmeing, T.M.; Moore, P.B.; Steitz, T.A. The Kink-Turn: A New RNA Secondary Structure Motif. *EMBO J.* **2001**, *20*, 4214–4221. [CrossRef]

32. Edwards, A.L.; Reyes, F.E.; Héroux, A.; Batey, R.T. Structural Basis for Recognition of S-Adenosylhomocysteine by Riboswitches. *RNA* **2010**, *16*, 2144–2155. [CrossRef]

33. Aalberts, D.P.; Hodas, N.O. Asymmetry in RNA Pseudoknots: Observation and Theory. *Nucleic Acids Res.* **2005**, *33*, 2210–2214. [CrossRef] [PubMed]

34. Maehigashi, T.; Dunkle, J.A.; Miles, S.J.; Dunham, C.M. Structural Insights into +1 Frameshifting Promoted by Expanded or Modification-Deficient Anticodon Stem Loops. *Proc. Natl. Acad. Sci. USA* **2014**, *111*, 12740–12745. [CrossRef] [PubMed]

35. Xu, W.; Wongsa, A.; Lee, J.; Shang, L.; Cannone, J.J.; Gutell, R.R. RNA2DMap: A Visual Exploration Tool of the Information in RNA's Higher-Order Structure. In Proceedings of the 2011 IEEE International Conference on Bioinformatics and Biomedicine, BIBM 2011, Atlanta, GA, USA, 12–15 November 2011; pp. 613–617. [CrossRef]

36. Borovinskaya, M.A.; Pai, R.D.; Zhang, W.; Schuwirth, B.S.; Holton, J.M.; Hirokawa, G.; Kaji, H.; Kaji, A.; Cate, J.H.D. Structural Basis for Aminoglycoside Inhibition of Bacterial Ribosome Recycling. *Nat. Struct. Mol. Biol.* **2007**, *14*, 727–732. [CrossRef] [PubMed]

37. Maguire, B.A.; Zimmermann, R.A. The Ribosome in Focus. *Cell* **2001**, *104*, 813–816. [CrossRef]

38. Jenner, L.; Starosta, A.L.; Terry, D.S.; Mikolajka, A.; Filonava, L.; Yusupov, M.; Blanchard, S.C.; Wilson, D.N.; Yusupova, G. Structural Basis for Potent Inhibitory Activity of the Antibiotic Tigecycline during Protein Synthesis. *Proc. Natl. Acad. Sci. USA* **2013**, *110*, 3812–3816. [CrossRef]

39. Ban, N.; Nissen, P.; Hansen, J.; Moore, P.B.; Steitz, T.A. The Complete Atomic Structure of the Large Ribosomal Subunit at 2.4 Å Resolution. *Science* **2000**, *289*, 905–920. [CrossRef]

Article

Noncovalent Bonds, Spectral and Thermal Properties of Substituted Thiazolo[2,3-b][1,3]thiazinium Triiodides

Irina Yushina *, Natalya Tarasova, Dmitry Kim, Vladimir Sharutin and Ekaterina Bartashevich

South Ural State University (National Research University) 76, Lenin Prospect, Chelyabinsk 454080, Russia;
tarasovanm@susu.ru (N.T.); kimdg@susu.ru (D.K.); sharutinvv@susu.ru (V.S.); bartashevichev@susu.ru (E.B.)
* Correspondence: iushinaid@susu.ru

Received: 31 August 2019; Accepted: 23 September 2019; Published: 28 September 2019

Abstract: The interrelation between noncovalent bonds and physicochemical properties is in the spotlight due to the practical aspects in the field of crystalline material design. Such study requires a number of similar substances in order to reveal the effect of structural features on observed properties. For this reason, we analyzed a series of three substituted thiazolo[2,3-b][1,3]thiazinium triiodides synthesized by an iodocyclization reaction. They have been characterized with the use of X-ray diffraction, Raman spectroscopy, and thermal analysis. Various types of noncovalent interactions have been considered, and an S . . . I chalcogen bond type has been confirmed using the electronic criterion based on the calculated electron density and electrostatic potential. The involvement of triiodide anions in the I . . . I halogen and S . . . I chalcogen bonding is reflected in the Raman spectroscopic properties of the I–I bonds: identical bond lengths demonstrate different wave numbers of symmetric triiodide vibration and different values of electron density at bond critical points. Chalcogen and halogen bonds formed by the terminal iodine atom of triiodide anion and numerous cation . . . cation pairwise interactions can serve as one of the reasons for increased thermal stability and retention of iodine in the melt under heating.

Keywords: chalcogen bond; halogen bond; triiodide anion; Raman spectroscopy; thermal analysis; thiazolo[2,3-b][1,3]thiazinium salts

1. Introduction

A great number of relatively strong noncovalent interactions in N- and S-heterocyclic polyiodides, such as halogen and chalcogen bonds, can not only promote the stability and diversity of possible polyiodide organization, but help in the organization of anion transport, chiral synthesis, and organocatalysis [1]. For electrostatically driven noncovalent bonds [2], the electrophilic site—the electron deficient region on the extension of the covalent bond of one atom—is orientated towards the nucleophilic fragment in another molecule. The International Union of Pure and Applied Chemistry (IUPAC) definition of halogen [3] and chalcogen bonds [4] is based on a series of quantitative features from structural, theoretical and spectral points of view. The development of noncovalent bond characterization in crystals includes the analysis of distribution features of the electrostatic potential [5], the quantum theory of atoms in molecules (QTAIM) methodology [6], electron localization function (ELF) [7], Laplacian of electron density [8], facilities of reduced density gradient (RDG) and analysis using the noncovalent interactions (NCI) method [9].

Variation of the N- and S-containing heterocyclic cations is useful for the investigation of bonding properties in organic polyiodide crystals. The chemistry of polyhalides has attracted particular attention due to the vast diversity of applicable properties related to the ability of halogens to form different types of contacts [10–15]. Polyiodides represent an absolute majority among polyhalide structures [16–21].

The interest in polyiodide derivatives is based on their enormous structural diversity of packing motives resulting in the formation of anionic ribbons, layers, channels and sheets [22–24]. The intense development of semiconductor [25–28] and nonlinear optical materials [29] based on polyiodides [30] necessitates obtaining and comprehensively investigating new molecular and crystal structures with specific properties. Structural investigation of the polarization effect or charge transfer in ionic complexes of thioamides with iodine is quite useful for explaning of the activity of some anti-thyroid drugs and antibiotics in vivo [31–36]. In all cases, the specific moiety responsible for biological activity is the released diiodide. The possibility of binding and releasing iodine from complex polyiodide structures finds its practical application in water disinfection [37] and the sorption of radioactive isotopes of iodine, such as I-131, and radiolabeling in vivo [38]. The particular significance of $I^{3-}/I^-·I_2$ interconversion is reflected in the tasks of the novel design of ionic liquids and iodine-containing solid electrolytes for dye-sensitized solar cells (DSSC) [39].

The use of such heterocyclic systems as thiazolo[2,3-a]isoquinolinium cations [40] is very promising in this regard. Syntheses of the related systems, such as 2*H*-benzo[4,5]thiazolo[2,3-b][1,3] thiazin-5-ium, are superficially described in the literature [41], and only a few examples of 3,5,6,7- tetrahydro-2*H*-thiazolo[2,3-b][1,3]thiazin-4-ium synthesis and transformation have been described [42–44] in the context of their biological activity.

The determination of crystal structure in polyiodide systems is highly necessary because of the ability of polyiodides to form mixtures of different compositions, to absorb excess iodine in a precipitate, and to form a melt or release iodine under storage conditions. Raman spectroscopy has earned a reputation as an extremely sensitive method [45] capable of distinguishing a variety of bonding peculiarities in the polyiodide anion [46]. It reliably characterizes polyiodides with various organic or inorganic cations, revealing typical structural units, such as the triiodide anion (100–120 cm^{-1}), bound iodine (140–180 cm^{-1}), and the pentaiodide anion (140–160 cm^{-1}) [30]. The role of thermal analysis is very important in the field of iodophoric materials and DSSC device development, where the questions of thermal stability, features of source decomposition, and iodine release are of particular importance [47].

Sometimes, for the S … I interactions in N- and S-heterocyclic polyiodides, the main question is which atom delivers the electrophilic site for bonding. If the I atom acts as the electrophilic site provider, we can conclude that this is a halogen bond. If the S atom delivers the electrophilic site, then it is a chalcogen bond. Thus, the relatively strong and charge-assisted S … I interactions in organic polyiodide crystals can be interpreted as either halogen or chalcogen bonds depending on the electron density distribution and mutual orientation of their electrophilic and nucleophilic sites. In order to figure out how the polarization effect or charge redistribution are directed, it is important to understand which atom donates electrons and which one delivers the electrophilic site for bonding. The categorization of chalcogen and halogen bonds in debatable cases can be performed using previously suggested [48] electronic criterion: "along the line between bound atoms, the minimum of electrostatic potential is always located from the side of nucleophilic site; the minimum of electron density is closer to electrophilic site provider".

Our research is focused on a series of thiazolo[2,3-b][1,3]thiazinium triiodides and on the study of how the spectral properties and thermal stability of these crystals are influenced by the type and features of the S … I noncovalent bond.

2. Materials and Methods

2.1. Synthesis

Thiazolo[2,3-b][1,3]thiazin-5-ium cations of compounds 2a–c (Figure 1) are formed via a halocyclization reaction by the action of the iodine molecule on an unsaturated bond of S-allylic or S-butenylic substituents. The compounds 3-iodo-4-phenyl-3,4-dihydro-2*H*-benzo[4,5]thiazolo[2,3-b] [1,3]thiazin-5-ium (2a) and 4-(iodomethyl)-3,4-dihydro-2*H*-benzo[4,5]thiazolo[2,3-b][1,3]thiazin-5-ium triiodides (2b) were obtained for the first time. Compounds 2a,b with aromatic systems in their structures crystallized faster than triiodide 2c with a partially unsaturated cation. High quality crystals of compound 2c were obtained in the mixture with a corresponding monoiodide only after recrystallization.

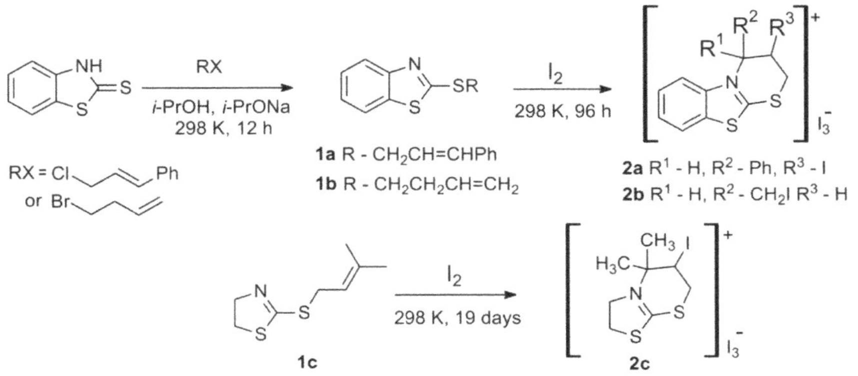

Figure 1. Iodocyclization of compounds 1a–c resulting in corresponding triiodides 2a–c.

The iodine solution was prepared as follows: anhydrous dioxane and CH_2Cl_2 were refluxed for half an hour with solid iodine. The solution was decanted from an excess of iodine and cooled. Te concentration of iodine was determined by titration with sodium thiosulphate solution in three replicates. The resulting concentration of iodine in dioxane was 73.9 ± 0.05 mg/mL, and 50.5 ± 0.05 mg/mL in CH_2Cl_2.

Benzothiazole-2-thione (2 g; 12 mmol) was dissolved in 20 mL i-PrOH with 12 mmol i-PrONa; then, 12 mmol alkenyl bromide (cinnamyl chloride or 4-bromobut-1-ene) was added dropwise. The mixture was stirred for 12 h at 298 K and filtered. The solution was diluted with 50 mL of water and extracted with three portions of CH_2Cl_2 (50 mL, 25 mL, and 25 mL). The extracts were combined, washed with 20 mL 5% H_2SO_4, 15 mL distilled water, and dried over $CaCl_2$. The solvent was evaporated under reduced pressure, giving compounds 1a,b in the form of yellow oil.

The compound 2-(cinnamylsulfanyl)benzo[d]thiazole (1a) was analyzed by ^{1}H NMR (CDCl$_3$, ppm): 7.90 (d, J = 8.2 Hz, H$_{Ar}$); 7.74 (d, J = 7.9 Hz, H$_{Ar}$); 7.44–7.38 (m, H$_{Ar}$); 7.38–7.32 (m, 2H$_{Ar}$); 7.31–7.26 (m, H$_{Ar}$); 7.25–7.20 (m, H$_{Ar}$); 6.70 (d, J = 15.7 Hz, =CH–); 6.37 (dt, J = 15.0, 7.3 Hz, =CH–); 4.17 (2H, d, J = 7.3 Hz, –CH$_2$–). It was also analyzed by ^{13}C NMR (dimethyl sulfoxide-*d6* (DMSO-*d6*), ppm): 166.21; 153.06; 136.31; 135.25; 134.27; 128.54 (2C); 127.87; 126.45 (2C); 126.04; 124.26; 123.45; 121.51; 120.96; 36.06. Elemental analysis calc. (%) for $C_{16}H_{13}NS_2$: C, 67.81; H, 4.62; N, 4.94; found: C, 67.60; H, 4.71; N, 4.88.

The compound 2-(butenylsulfanyl)benzo[d]thiazole (1b) was analyzed by ^{1}H NMR: (CDCl$_3$, ppm): 7.86 (m, H$_{Ar}$); 7.75 (m, H$_{Ar}$); 7.41 (m, H$_{Ar}$); 7.29 (m, H$_{Ar}$); 5.89 (m, =CH); 5.13 (2H, m, =CH$_2$); 3.43 (2H, t, J = 7.3 Hz, SCH$_2$); 2.59 (2H, m, –CH$_2$–). ^{13}C NMR (DMSO-*d6*, ppm): 166.81; 153.28; 135.67; 135.18; 125.97; 124.14; 121.46; 120.90; 116.92; 33.31; 32.71. Elemental analysis calc. (%) for $C_{11}H_{11}NS_2$: C, 59.69; H, 5.01; N, 6.33; found: C 59.60, H 5.15, N 6.35. The compound 2-(3-methylbutenyl)sulfanyl-4,5-dihydrothiazole (1c) was obtained as described elsewhere [49].

For the compound 3-iodo-4-phenyl-3,4-dihydro-2*H*-benzo[4,5]thiazolo[2,3-b][1,3]thiazin-5-ium triiodide (2a), compound 1a (1 mmol) was dissolved in 1 mL anhydrous CH_2Cl_2, and 7.11 mL of a solution of iodine (50.5 ± 0.05 mg/mL) was added under stirring. The resulting mixture was kept at room temperature for 4 days. Dark brown crystals were filtered and dried in vacuo (yield 0.46 g, 98%). It was analyzed by ^{1}H NMR (DMSO-*d6*, ppm): 8.46–8.42 (m, H$_{Ar}$); 7.90–7.84 (m, H$_{Ar}$); 7.78–7.69 (m, 2H$_{Ar}$); 7.50–7.43 (m, H$_{Ar}$); 7.38–7.24 (m, H$_{Ar}$); 7.01 (1H, s, 4-H); 5.69 (1H, dd, J = 5.9, 3.0 Hz, 3-H); 3.70

(1H, dd, J = 14.6, 3.4 Hz, 2-H); 3.40 (1H, dd, J = 14.6, 2.8 Hz, 2-H) and ^{13}C NMR (DMSO-*d6*, ppm): 175.96; 140.67; 135.42; 129.60 (2C); 129.50; 129.18; 128.24; 127.43; 126.15 (2C); 124.53; 114.67; 66.33; 33.78; 17.19. Elemental analysis calc. (%) for $C_{16}H_{13}I_4NS_2$: C, 24.29; H, 1.66; N, 1.77; found: C, 24.31; H, 1.62; N, 1.83.

For the compounds 4-(iodomethyl)-3,4-dihydro-2*H*-benzo[4,5]thiazolo[2,3-b][1,3]thiazin-5-ium triiodide (2b) and 6-iodo-5,5-dimethyl-3,5,6,7-tetrahydro-2*H*-thiazolo[2,3-b][1,3]thiazin-4-ium triiodide (2c): compounds 1b,c (1 mmol) were dissolved in 0.8 mL anhydrous dioxane, and 7.35 mL of a solution of iodine (73.9 ± 0.05 mg/mL) was added under stirring. The resulting mixture was kept at room temperature for 19 days (2b) and 8 days (2c).

For 4-(iodomethyl)-3,4-dihydro-2*H*-benzo[4,5]thiazolo[2,3-b][1,3]thiazin-5-ium triiodide (2b): dark brown crystals were filtered and dried in vacuo (yield 0.6 g, 91%); analysis by ^{1}H NMR (DMSO-*d6*, ppm): 8.34 (d, J = 7.7, H$_{Ar}$); 8.12 (d, J = 8.5 Hz, H$_{Ar}$); 7.89 (t, J = 7.9 Hz, H$_{Ar}$); 7.72 (t, J = 7.7 Hz, H$_{Ar}$); 5.58–5.51 (1H, m, 5-CH); 3.76–3.66 (2H, m, CH$_2$I); 3.65–3.57 (2H, m, 3-CH$_2$); 3.05–2.96 (1H, m, 4-CH$_2$); 2.49–2.41 (1H, m, 4-CH$_2$) and ^{13}C NMR (DMSO-*d6*, ppm):: 173.63; 136.67; 126.58; 125.55; 124.89; 121.83; 119.48; 112.87; 53.15; 30.11; 21.41. Elemental analysis calc. (%) for $C_{11}H_{11}I_4NS_2$: C, 18.12; H, 1.52; N, 1.92; found: C, 18.10; H, 1.60; N, 1.89.

For 6-iodo-5,5-dimethyl-3,5,6,7-tetrahydro-2*H*-thiazolo[2,3-b][1,3]thiazin-4-ium triiodide (2c): the solvent was decanted and dark brown crystals with a matte surface (0.43 g) were isolated and dried in vacuo. A portion of the crystals (0.23 g) was recrystallized from a 5 mL mixture of i-PrOH – DMF (N,N-Dimethylformamide) (3:2). The mixture of small brown and yellow crystals (0.15 g) was filtered and dried in vacuo (Figure S1, see Supplementary Materials). The crystal structure of the brown prismatic crystals (2c) was determined using single crystal X-ray diffraction. The impurity of the yellow prismatic crystals, which should presumably relate to the corresponding monoiodide, was separated manually under a microscope, but still did not prove suitable for X-ray diffraction experiments (Figure S1).

2.2. X-ray Diffraction Refinement

X-ray diffraction study of single crystals 2a–c was carried out with a Bruker D8 QUEST diffractometer (Mo-Kα radiation, λ = 0.71073 Å, graphite monochromator). Collection, handling of data and refinement of the unit cell parameters, as well as accounting for absorption, were carried out using the SMART and SAINT-Plus programs [50]. All calculations were performed using SHELXTL/PC [51] and OLEX2 [52] software. The structure was solved by direct method and refined by the method of least squares in the anisotropic approximation for non-hydrogen atoms. Crystal data and structure refinement parameters can be found in Table 1. A full list of bond lengths and valence angles are shown in Tables S1 and S2. Full crystallographic data for the compounds can be obtained free of charge from The Cambridge Crystallographic Data Centre (CCDC) (1829960 (2a), 1589820 (2b), 1589819 (2c)).

2.3. Sample Characterization

The ^{1}H NMR and ^{13}C NMR spectra were recorded on a Bruker Avance-500 500 and 126 MHz apparatus (tetramethylsilane as an internal standard). The elemental compositions were determined with a Carlo Erba CHNS-O EA 1108 analyzer. Raman spectra were obtained with a NTEGRA Spectra spectrometer using a 632.8 nm line of the He–Ne$^+$ laser for spectra excitation. The laser power on the sample surface was about 0.2 µW. We had to take into consideration that due to decreased thermal stability of higher polyiodides [53], the laser power should be thoroughly controlled in order to avoid decomposition under laser explosion. The thermal analysis data have been obtained in the temperature range of 25–700 °C using Netzsch STA F1 equipment at 10 K/min^{-1} heating rates in corundum crucibles in an air atmosphere. The masses of the analyzed samples were 2.0–2.2 mg.

Table 1. Crystal data and structure refinement for triiodides 2a–c.

Structure	2a	2b	2c
Empirical formula	$C_{16}H_{13}NS_2I_4$,	$C_{11}H_{11}NS_2I_4$,	$C_8H_{13}NS_2I_4$
Temperature (K)	293	293	296.15
Crystal system	triclinic	triclinic	monoclinic
Space group	P-1	P-1	P21/n
a (Å)	9.775(7)	7.913(5)	12.1447(7)
b (Å)	9.868(6)	7.994(5)	10.2762(5)
c (Å)	12.890(8)	14.883(11)	13.9862(7)
α (°)	96.416(19)	101.68(3)	90
β (°)	98.49(2)	96.33(3)	108.053(2)
γ (°)	117.72(4)	101.22(3)	90
Volume (Å^3)	1065.24	893.29	1659.57(15)
Z	2	2	4
Density (g/cm^3)	2.466	2.710	2.781
μ, (mm^{-1})	6.045	7.195	7.74
F(000)	720.0	653.4	1242.9
Crystal size (mm)	$0.25 \times 0.14 \times 0.11$	$0.77 \times 0.27 \times 0.22$	$0.47 \times 0.29 \times 0.17$
2θ range of data collection (deg)	5.9 to 59.22	5.66 to 66.48	6.12 to 79.2
Range of refraction indices	$-13 \leq h \leq 13$, $-13 \leq k \leq 13$, $-17 \leq l \leq 17$	$-12 \leq h \leq 12$, $-12 \leq k \leq 12$, $-22 \leq l \leq 22$	$-21 \leq h \leq 21$, $-18 \leq k \leq 18$, $-24 \leq l \leq 25$
Measured reflections	47057	52396	81480
Independent reflections	5968 ($R_{int} = 0.0346$, $R_{sigma} = 0.0186$	6839 ($R_{int} = 0.0357$, $R_{sigma} = 0.0208$)	9984 ($R_{int} = 0.0432$, $R_{sigma} = 0.0315$)
Refinement variables	211	165	142
Goodness-of-fit on F^2	1.039	1.049	1.104
R factors for $F^2 > 2\sigma(F^2)$	$R_1 = 0.0283$, $wR_2 = 0.0623$	$R_1 = 0.0450$, $wR_2 = 0.1007$	$R_1 = 0.0543$, $wR_2 = 0.117$
R factors for all reflections	$R_1 = 0.0431$ $wR_2 = 0.0683$	$R_1 = 0.0611$ $wR_2 = 0.1098$	$R_1 = 0.1107$ $wR_2 = 0.1171$
Residual electron density (min/max) (e/Å^3)	0.61/−1.29	1.98/−2.24	3.13/−3.71

2.4. Theoretical Calculations

Periodic Kohn–Sham calculations were performed in the CRYSTAL17 program package [54], employing the B3LYP exchange-correlation functional. K-point sampling was done using a Monkhorst–Pack grid of $8 \times 8 \times 8$. The modified DZVP basis set was used for iodine atoms [55], and the 6-31G** Gaussian type basis sets were used for the C, H, N and S atoms from [56]. Calculations of the electron localization function (ELF) [7] distribution in the planes, in which the S...I interactions lay, were performed using the TOPOND program [57]. The values of the electron density (ED) and electrostatic potential (ESP) along the line between the S and I atoms were derived due to the additional output procedure in TOPOND approximating the interatomic line by 200 points.

3. Results and Discussion

3.1. Structural Characterization of Compounds 2a–c

The crystal structures of compounds 2a,b belong to the triclinic crystal system. The crystallographic cell consists of two ionic pairs of 2H-benzo[4,5]thiazolo[2,3-b][1,3]thiazin-5-ium cations with triiodide anions. Compound 2c crystallizes in a monoclinic lattice with four ionic pairs of 3,5,6,7-tetrahydro-2H-thiazolo[2,3-b][1,3]thiazin-4-ium cations with triiodide anions (Figure 2). The summary of geometric parameters of the S...I interactions and bond lengths within triiodide anions can be found in Table 2.

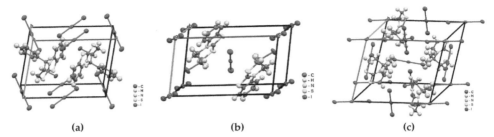

| | (a) | (b) | (c) |

Figure 2. Crystal packing of 2a (**a**); 2b (**b**) and 2c (**c**) with thermal ellipsoids of 50% probability, view along the b crystallographic axis.

Table 2. Experimental geometric parameters of S . . . I interactions in 2a–c crystal structures.

	Distance S . . . I, Å	C–S . . . I Angle, °	S . . . I–I Angle, °	I–I in I_3^-, Å
2a	S2 . . . I2: 3.707	C1–S2 . . . I2: 142.27	S2 . . . I2–I4: 110.54	2.917; 2.917
2b	S2 . . . I2: 3.775	C1–S2 . . . I2: 161.54	S2 . . . I2–I1: 110.53;	I_3^- (1): 2.912; 2.912
	S2b . . . I2: 3.778	C10–S2b . . . I2: 153.69	S2b . . . I2–I1: 32.93	I_3^- (2): 2.931; 2.931
2c	S4 . . . I3: 3.910	C9–S4 . . . I3: 165.72	S4 . . . I3–I2: 96.39	I_3^- (1): 2.904; 2.936
	S3 . . . I8: 3.699	C9–S3 . . . I8: 166.14	S3 . . . I8–I7: 96.41	
	S2 . . . I1: 3.902	C1–S2 . . . I1: 172.07	S2 . . . I1–I2: 77.78	I_3^- (2): 2.931; 2.933
	S1 . . . I6: 3.734	C1–S1 . . . I6: 165.23	S1 . . . I6–I7: 96.74	

3.2. Noncovalent Bonds Formed by Triiodide Anions: Electron Density Calculations

In crystal 2a, there is a typical halogen bond between the terminal iodine atom I4 of the triiodide anion and a covalently bound iodine atom I5 as part of the CH–I fragment. Both the I4 and I2 atoms of the triiodide anion in crystal 2a form the chalcogen bonds with S atoms of the heterocyclic cation. In the triiodide anion of crystal 2b, both terminal atoms form two S . . . I interactions with almost identical distances, but they differ in their C–S . . . I and S . . . I–I angle values (Table 2); the central I1 and terminal I2 atoms are also involved in the I1 . . . H9 and I2 . . . H9 hydrogen bonds. Crystal structure 2c represents a case of a typical chalcogen bond where the C8 . . . S2–I3 angle tends to 180° and S2–I3–I1 is nearly a right angle, so that the mutual orientation of the electrophilic site of S2 and the nucleophilic region of I3 fits in the best way. Note that the terminal I atoms in the triiodide anions of 2c are involved in multiple noncovalent interactions, among which the most significant are chalcogen bonds S4 . . . I3, S3 . . . I8, S1 . . . I6, S2 . . . I1 and halogen bond I3 . . . I5 with iodine atom I5 in the organic cation.

The representation of regions with a concentration and depletion of electron pairs in molecules and crystals can be done using the electron localization function (ELF) [7]. This function is dimensionless and is normalized from 0 to 1, where the value 0.5 corresponds to the case of uniformly distributed one-electron gas. It is particularly interesting in the case of halogen atoms, as it is known that their electron concentration regions form the belt in the equatorial part of an atom [58]. ELF for I atoms in the triiodide anions (Figure 3) clearly reveals the regions of electron accumulation (shown in orange) and electrophilic sites—regions of electron depletion on the extension of the I–I bond. However, for the bound S atom, the electrophilic site region is not always clearly seen in ELF, because it can be camouflaged by the electron concentration region that give a pronounced projection of high ELF values onto the considered plane of atomic interactions. This fact complicates the identification of the orientation of the bound S and I atoms. Accordingly, in such a case, we have applied the electronic criterion [48] that allows us to identify the atom (S or I) that has provided the electrophilic site and defined the name of interaction; in our case, it is either a halogen or chalcogen bond. The examples of the main considered S . . . I interactions in the studied crystals are presented in Figures 3 and 4.

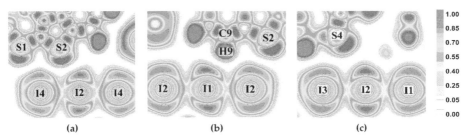

Figure 3. Electron localization function (ELF) distribution in S ... I motives in the structures (**a**) 2a (plane I4—I2 ... S2); (**b**) 2b (plane I1–I2 ... S2); and (**c**) 2c (plane I3–I2 ... S4).

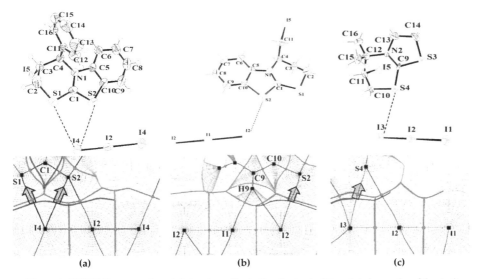

Figure 4. ORTEP diagrams of crystal structures illustrating the typical S ... I chalcogen bond (on top); on bottom: gradient fields, atomic basins and bond critical points of electron density (red lines and points), electrostatic potential distributions (blue lines) for typical chalcogen bonds (**a**) 2a (plane I4—I2 ... S2); (**b**) 2b (plane I2–I1 ... S2); and (**c**) 2c (plane I3–I2 ... S4).

Let us consider the S ... I and other interactions in the studied crystals from the calculated electron density and electrostatic potential point of view as this approach has already been proven to be demonstrative [59,60]. Figure 4 illustrates the examples of S ... I chalcogen bonds in the considered crystals as well as the verification of non-covalent bonding type based on the electronic criteria. Within QTAIM [6], a zero-flux condition of the gradient vector field of electron density defines the boundaries of chemically bonded atoms, while basin boundaries in the electrostatic potential demarcate electrically neutral atomic fragments [61]. The map of the gradient field of electron density (Figure 4) allows us to recognize the boundaries of atomic basins. If this map is superimposed on the map of electrostatic potential distribution, it can be seen that the boundaries of electrically neutral atoms located using electrostatic potential do not coincide with the boundaries of atomic basins in electron density. This means that the fraction of electrons formally belonging to I atoms are attracted to S nuclei along the highlighted directions. This is the general feature of chalcogen bonding. The images in Figure 4 are accompanied by the bond critical points (bcp) with values of electron density given in Table S3.

As a simplification, the order of electron density and electrostatic potential minima along the S . . . I line can give a clue about the localization of the electrophilic site for typical and unobvious cases of halogen/chalcogen bonding. In this case, the formulation of previously suggested electronic criterion becomes more demonstrative. For the S1 . . . I4 (2a), S2 . . . I2 (2b) and S4 . . . I3 (2c) bonds in Figure 5, the minima of electrostatic potential (ESP) are located at the side of the I atoms donating electrons; the minima of electron density (ED) are placed closer to the S atoms which deliver their electrophilic site for noncovalent bonding. The arrows in Figure 5 show the direction in which the electron density of the I atom is attracted to the electrophilic site of the S atom. Thus, for all the cases of illustrated I . . . S noncovalent bonds in 2a–2c, we ascribe such interactions to a chalcogen bonding.

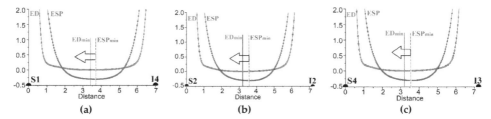

Figure 5. One-dimensional distributions of electron density (ED) and electrostatic potential (ESP) functions (a.u.) along the S . . . I chalcogen bonds in 2a (**a**), 2b (**b**) and 2c (**c**) illustrating that the positions of electron density minimum (ED$_{min}$) are closer to the S atoms; interatomic distances are in Bohr.

3.3. Raman Spectroscopy Data

Experimental Raman spectra of the studied crystals are presented in Figure 6. On the spectrum of crystal 2c, we clearly see two overlapping bands corresponding to the two symmetrically independent triiodide anions in the crystallographic cell. As both of them are symmetric, we see only the band of symmetric stretching vibration of the triiodide anion at 112.1 cm^{-1} and 113.7 cm^{-1} and do not see the band of antisymmetric vibration, usually observed in the range of 120–140 cm^{-1} [30].

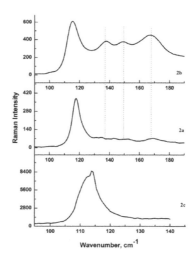

Figure 6. Experimental Raman spectra of crystals 2a–c in the low-frequency region.

The spectra of crystals 2a and 2b are very similar and have weak bands at 138 cm^{-1}, 149 cm^{-1} and 168 cm^{-1}, which are not attributed to the anionic vibrations and relate to libration and deformational vibrations of the tricyclic dihydrobenzothiazolo[2,3-b]thiazin-5-ium system. The intense triiodide vibration bands are located at 118 cm^{-1} (2a) and 115 cm^{-1} (2b).

The correlation between the I–I bond lengths in the triiodide anion and the observed experimental wavenumbers of symmetric vibration is not so clear and evident as in case of iodine complexes [62,63]. For example, one of the triiodide anions in the crystal 2c has the same bond length as in 2a (2.9170 Å), but its wavenumber is much lower: 113 cm^{-1} (2c) and 118 cm^{-1} (2a), respectively. Such a spectral difference may reveal the effects that are beyond the geometric contribution on the level of bond length analysis and may be due to the effect of noncovalent interactions and crystal surrounding. Thus, as the isolated triiodide in 2c demonstrates a lower wavenumber, it has weaker I–I bonding within the triiodide anion in comparison to the anion in crystal 2a, demonstrating the band at 118 cm^{-1}.

Such features of bonding strength in triiodide anions according to spectral data are also reflected in quantum-topological analysis of electron density. The electron density value at the bond critical point of the I–I bond in the triiodide anion is slightly higher in crystal 2a (0.0433 a.u.) than in crystal 2c (0.0419 a.u.), although the calculated equilibrium distances in these structures are identical (Table 2).

This fact is in agreement with the experimental spectral data and electron density values at the I–I bond critical point. Thus, the combination of experimental Raman characteristics together with calculated electron density descriptors can enrich the understanding of bond properties within the triiodide anion beyond the geometric approach based on bond length analysis.

3.4. Thermal Analysis Data

Thermoanalytical curves of TG-DSC analysis for crystals 2a–c are presented in Figure 7. The summary of the decomposition characteristics can be found in Table 3. All samples demonstrate multistep decomposition and rather high thermal stability for polyiodides; decomposition in all cases starts above 150 °C.

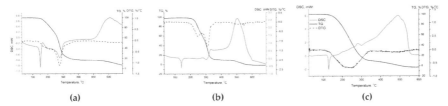

| | (a) | (b) | (c) |

Figure 7. Thermogravimetric (TG), differential scanning calorimetry (DSC) and differential thermogravimetric (DTG) curves of crystals: 2a (**a**), 2b (**b**) and 2c (**c**).

Table 3. Thermal decomposition features of crystals 2a–c.

	2a	2b	2c
Melting Point (°C)	141	127	122.5
Decomposition start (°C)	162*smooth	222	160
DTG peak 1 (°C)	180–210	249	215
DTG peak 2 (°C)	277	310	252
Mass loss, peak 1 and 2 (%)	83.4	89.9	90.3

We can attempt to compare thermal decomposition and melting temperatures in the series of heterocyclic derivatives with the thiazolo-thiazinium fragment and with the same anion composition in order to try and find some trends due to differences in noncovalent interactions in the analyzed structures. A comparison of the data listed in Tables 2 and 3 shows that melting points decrease in the row 2a > 2b > 2c (Table 1). The structural complexity of cations reduces in the order 2a – 2b – 2c from the tricyclic system with the voluminous phenyl substituent (2a) to the same tricyclic system but with the small –CH$_2$I substituent (2b) to the bicyclic thiazolo-thiazinium system (2c), leading to a lower melting point for structure 2c.

All studied triiodides tend to decompose only in the melt (no mass loss is observed before the melting point), although here the universal trend is not so evident: structure 2b demonstrates remarkably high thermal stability (222 °C, almost a 100 °C stability range of the melt), which is significantly higher than that for the other two structures. This fact may be due to the existence of four S ... I interactions involving both the terminal iodine atoms of triiodide anion in 2b, organized in such a way that the anion is trapped into the cage of four neighboring organic cations (Figure 3b). This fact can lead to a higher retention of iodine in the melt, as typically the first decomposition stage of polyiodides usually includes the loss of I$_2$ from the triiodide anion [64,65]. Similar conclusions concerning melting and decomposition of crystals with C–Se ... O/N chalcogen bonds were found in a recent study [66].

The crystal structures of compounds 2a and 2c are similar to the zigzag-like location of triiodide anions, while the crystal structure of 2b demonstrates layered packing of organic cations, as the benzothiazolo-thiazinuim cycle is flatter than the corresponding thiazolo-thiazinium and especially than phenyl-substituted 2a. Such layered organization in crystal 2b can lead to more effective capsulation of triiodide anions in the cage of the cations, which can be one of possible reasons for comparatively higher retention of iodine in the melt in the row of triiodides 2a – 2b – 2c. Based on quantum-topological analysis of the electron density properties (Tables S3 and S4) and thermal analysis data, we can make some inferences. Firstly, both triiodides in structure 2b are relatively weakly bound with crystalline surrounding; the only rather strong chalcogen bond S ... I (3.7208 Å, the value of electron density at the bond critical point (r_{bcp}) = 0.008 a.u.) dominates. Secondly, in structure 2a, we observe a moderate number of bond critical points but only for one symmetrically independent triiodide anion. The other is weakly bound to the crystal surrounding. Thirdly, in structure 2c, the numerous weak I ... H interactions are formed with each of the triiodide anions; one relatively short and strong I ... H hydrogen bond (2.8940 Å, (r_{bcp}) = 0.0114 a.u.) stands out from all others. Chalcogen S ... I and halogen I ... I bonds between the cation and anion are also observed. In general, the 2a > 2b > 2c melting point order is reversed for the overall density in the bond critical points of cation ... cation (H ... H, C ... I, S ... H, C ... C) interactions.

4. Conclusions

The synthesized crystal structures of 2*H*-benzo[4,5]thiazolo[2,3-b][1,3]thiazin-5-ium and 3,5,6,7-tetrahydro-2*H*-thiazolo[2,3-b][1,3]thiazin-4-ium triiodides were characterized using single crystal X-ray diffraction, Raman spectroscopy, and thermal analysis. The type of noncovalent bonds were described on the basis of periodic quantum-chemical calculations revealing cation ... anion chalcogen S ... I bonds and halogen I ... I bonds. In the cases of ambiguous mutual orientation of noncovalently bound atoms, the electronic criterion of the disposition of electron density and electrostatic potential minima was used; all considered S ... I interactions were categorized as chalcogen bonds. Raman spectra and local properties of electron density allowed us to reveal the influence of noncovalent bonds on the properties of triiodide anions with equal bond length: the more bound iodine atoms was the reason for a stronger I–I bond within the triiodide anion. The thermal analysis data showed that the layered packing of benzothiazolo-thiazinium triiodide promoted effective capsulation of triiodide anions due to hydrogen and chalcogen bonding and might act as a stabilizing factor, providing comparatively higher thermal stability and iodine retention in the melt.

Supplementary Materials: The following are available online at http://www.mdpi.com/2073-4352/9/10/506/s1, Table S1: Bond lengths for triiodides 2a, 2b and 2c, Table S2: Bond angles for triiodides 2a, 2b and 2c, Figure S1: Crystal habit of triiodide 2c (ruby red) and corresponding monoiodide (light yellow) under 4× microscope magnification, Table S3: Local properties at bond critical points (bcp) of noncovalent interactions formed by triiodide anions in 2a, 2b and 2c crystals, Table S4: Local properties at bond critical points of noncovalent interactions formed by organic cations in 2a, 2b and 2c.

Author Contributions: Conceptualization, E.B.; Funding acquisition, D.K., V.S. and E.B.; Investigation, I.Y., N.T. and V.S.; Methodology, E.B.; Project administration, D.K. and E.B.; Resources, N.T.; Supervision, D.K. and E.B.; Validation, I.Y. and E.B.; Visualization, I.Y. and E.B.; Writing—original draft, I.Y., N.T.; Writing—review and editing, E.B.

Funding: The work was supported by The Government of Russian Federation decree №211, agreement № 02.A03.21.0011 and by the Ministry of Education and Science of the Russian Federation: grant 4.9665.2017/8.9 (N.T. and D.K.), grant 4.6151.2017/8.9 (V.S.) and 4.1157.2017/4.6 (I.Y. and E.B.).

Conflicts of Interest: The authors declare no conflict of interest.

References

1. Vogel, L.; Wonner, P.; Huber, S.M. Chalcogen Bonding: An Overview. *Angew. Chem.* **2019**, *58*, 1880–1891. [CrossRef] [PubMed]
2. Politzer, P.; Murray, J.S.; Clark, T. Halogen Bonding: An Electrostatically-Driven Highly Directional Noncovalent Interaction. *Phys. Chem. Chem. Phys.* **2010**, *12*, 7748–7757. [CrossRef] [PubMed]
3. Desiraju, G.R.; Ho, P.S.; Kloo, L.; Legon, A.C.; Marquardt, R.; Metrangolo, P.; Politzer, P.; Resnati, G.; Rissanen, K. Definition of the Halogen Bond (IUPAC Recommendations 2013). *Pure Appl. Chem.* **2013**, *85*, 1711–1713. [CrossRef]
4. Aakeroy, C.B.; Bryce, D.L.; Desiraju, G.R.; Frontera, A.; Legon, A.C.; Nicotra, F.; Rissanen, K.; Scheiner, S.; Terraneo, G.; Metrangolo, P.; et al. Definition of the chalcogen bond. *Iupac Recomm.* **2019**. [CrossRef]
5. Politzer, P.; Murray, J.S. An Overview of Strengths and Directionalities of Noncovalent Interactions: σ-Holes and ρ-Holes. *Crystals* **2019**, *9*, 165. [CrossRef]
6. Bader, R.F.W. *Atoms in Molecules: A Quantum Theory*; Clarendon Press: Oxford, UK, 1990; pp. 1–438.
7. Becke, A.D.; Edgecombe, K.E. A simple measure of electron localization in atomic and molecular systems. *J. Chem. Phys.* **1990**, *92*, 5397–5403. [CrossRef]
8. Tsirelson, V.G.; Zhou, P.F.; Tang, T.-H.; Bader, R.F.W. Topological definition of crystal structure: Determination of the bonded interactions in solid molecular chlorine. *Acta Crystallogr. Sect. A* **1995**, *A51*, 143–153. [CrossRef]
9. Johnson, E.R.; Keinan, S.; Mori-Sánchez, P.; Contreras-García, J.; Cohen, A.J.; Yang, W. Revealing Non-Covalent Interaction. *J. Am. Chem. Soc.* **2010**, *132*, 6498–6506. [CrossRef]
10. Adonin, S.A.; Bondarenko, M.A.; Novikov, A.S.; Abramov, P.A.; Plyusnin, P.E.; Sokolov, M.N.; Fedin, V.P. Halogen Bonding-assisted Assembly of Bromoantimonate (V) and Polybromide-bromoantimonate-based Frameworks. *CrystEngComm* **2019**, *2*, 850–856. [CrossRef]
11. Adonin, S.A.; Gorokh, I.D.; Abramov, P.A.; Novikov, A.S.; Korolkov, I.V.; Sokolov, M.N.; Fedin, V.P. Chlorobismuthates Trapping Dibromine: Formation of Two-Dimensional Supramolecular Polyhalide Networks with Br2 Linkers. *Eur. J. Inorg. Chem.* **2017**, 4925–4929. [CrossRef]
12. Adonin, S.A.; Bondarenko, M.A.; Abramov, P.A.; Novikov, A.S.; Plyusnin, P.E.; Sokolov, M.N.; Fedin, V.P. A Bromo- and polybromoantimonates (V): Structural and theoretical studies of hybrid halogen-rich halometalate frameworks. *Chem. Eur. J.* **2018**, *24*, 10165–10170. [CrossRef] [PubMed]
13. Kukkonen, E.; Malinen, H.; Haukka, M.; Konu, J. Reactivity of 4-Aminopyridine with Halogens and Interhalogens: Weak Interactions Supported Networks of 4-Aminopyridine and 4-Aminopyridinium. *Cryst. Growth Des.* **2019**, *19*, 2434–2445. [CrossRef]
14. Gross, M.M.; Manthiram, A. Long-Life Polysulfide-Polyhalide Batteries with a Mediator-Ion Solid Electrolyte. *ACS Appl. Energy Mater.* **2019**, *2*, 3445–3451. [CrossRef]
15. Sonnenberg, K.; Pröhm, P.; Schwarze, N.; Müller, C.; Beckers, H.; Riedel, S. Investigation of Large Polychloride Anions: [Cl11]−, [Cl12]2−, and [Cl13]−. *Angew. Chem. Int. Ed.* **2018**, *57*, 9136–9140. [CrossRef] [PubMed]
16. Savastano, M.; Martínez-Camarena, Á.; Bazzicalupi, C.; Delgado-Pinar, E.; Llinares, J.M.; Mariani, P.; Verdejo, B.; García-España, E.; Bianchi, A. Stabilization of Supramolecular Networks of Polyiodides with Protonated Small Tetra-azacyclophanes. *Inorganics* **2019**, *7*, 48. [CrossRef]
17. Yu, H.-L.; He, Y.-C.; Zhao, F.-H.; Wang, Y.; Wang, A.-N.; Hao, M.-G.; Si, Z.-S.; You, J. Synthesis, Structure and Properties of a New Polyiodide Compound with the 1D→3D Interdigitated Architecture. *Polyhedron* **2019**, *169*, 183–186. [CrossRef]
18. McDaniel, J.G.; Yethiraj, A. Grotthuss Transport of Iodide in EMIM/I3 Ionic Crystal. *J. Phys. Chem. B* **2018**, *122*, 250–257. [CrossRef] [PubMed]

19.   Shestimerova, T.A.; Bykov, M.A.; Wei, Z.; Dikarev, E.V.; Shevelkov, A.V. Crystal Structure and Two-level Supramolecular Organization of Glycinium Triiodide. *Russ. Chem. Bull.* **2019**, *68*, 1520–1524. [CrossRef]

20.   Shestimerova, T.A.; Golubev, N.A.; Yelavik, N.A.; Bykov, M.A.; Grigorieva, A.V.; Wei, Z.; Dikarev, E.V.; Shevelkov, A.V. Role of I2 Molecules and Weak Interactions in Supramolecular Assembling of Pseudo-Three-Dimensional Hybrid Bismuth Polyiodides: Synthesis, Structure, and Optical Properties of Phenylenediammonium Polyiodobismuthate(III). *Cryst. Growth Des.* **2018**, *18*, 2572–2578. [CrossRef]

21.   Savinkina, E.V.; Golubev, D.V.; Grigoriev, M.S. Synthesis, Characterization, and Crystal Structures of Iodides and Polyiodides of Scandium Complexes with Urea and Acetamide. *J. Coord. Chem.* **2019**, *72*, 347–357. [CrossRef]

22.   Ma, L.; Peng, H.; Lu, X.; Liu, L.; Shao, X. Building up 1-D, 2-D, and 3-D Polyiodide Frameworks by Finely Tuning the Size of Aryls on Ar-S-TTF in the Charge-Transfer (CT) Complexes of Ar-S-TTFs and Iodine. *Chin. J. Chem.* **2018**, *36*, 845–850. [CrossRef]

23.   Blake, A.J.; Li, W.-S.; Lippolis, V.; Schröder, M.; Devillanova, F.A.; Gould, R.O.; Parsons, S.; Radek, C. Template self-assembly of polyiodide networks. *Chem. Soc. Rev.* **1998**, *27*, 195–206. [CrossRef]

24.   Rybkovskiy, D.V.; Impellizzeri, A.; Obraztsova, E.D.; Ewels, C.P. Polyiodide Structures in Thin Single-walled Carbon Nanotubes: A large-scale Density-functional Study. *Carbon* **2019**, *142*, 123–130. [CrossRef]

25.   Li, H.-H.; Chen, Z.-R.; Cheng, L.-C.; Liu, J.-B.; Chen, X.-B.; Li, J.-Q. A New Hybrid Optical Semiconductor Based on Polymeric Iodoplumbate Co-Templated by Both Organic Cation and Polyiodide Anion. *Cryst. Growth Des.* **2008**, *8*, 4355–4358. [CrossRef]

26.   Wlaźlak, E.; Kalinowska-Tłuścik, E.J.; Nitek, W.; Klejna, S.; Mech, K.; Macyk, W.; Szaciłowski, K. Triiodide Organic Salts: Photoelectrochemistry at the Border between Insulators and Semiconductors. *Chemelectrochem* **2018**, *5*, 3486–3497. [CrossRef]

27.   Poręba, T.; Ernst, M.; Zimmer, D.; Macchi, P.; Casati, N. Pressure-Induced Polymerization and Electrical Conductivity of a Polyiodide. *Angew. Chem. Int. Ed.* **2019**, *58*, 6625–6629. [CrossRef]

28.   Starkholm, A.; Kloo, L.; Svensson, P.H. Polyiodide hybrid perovskites: A strategy to convert intrinsic 2D systems into 3D photovoltaic materials. *ACS Appl. Energy Mater.* **2019**, *2*, 477–485. [CrossRef]

29.   Yin, Z.; Wang, Q.-X.; Zeng, M.-H. Iodine Release and Recovery, Influence of Polyiodide Anions on Electrical Conductivity and Nonlinear Optical Activity in an Interdigitated and Interpenetrated Bipillared-Bilayer Metal−Organic Framework. *J. Am. Chem. Soc.* **2012**, *134*, 4857–4863. [CrossRef]

30.   Svensson, P.H.; Kloo, L. Synthesis, Structure, and Bonding in Polyiodide and Metal Iodide−Iodine Systems. *Chem. Rev.* **2003**, *103*, 1649–1684.

31.   Corban, G.J.; Hadjikakou, S.K.; Hadjiliadis, N.; Kubicki, M.; Tiekink, E.R.T.; Butler, I.S.; Drougas, E.; Kosmas, A.M. Synthesis, Structural Characterization, and Computational Studies of Novel Diiodine Adducts with the Heterocyclic Thioamides N-Methylbenzothiazole-2-thione and Benzimidazole-2-thione: Implications with the Mechanism of Action of Antithyroid Drugs. *Inorg. Chem.* **2005**, *44*, 8617–8627. [CrossRef]

32.   Roy, G.; Nethaji, M.; Mugesh, G. Interaction of anti-thyroid drugs with iodine: The isolation of two unusual ionic compounds derived from Se-methimazole. *Org. Biomol. Chem.* **2006**, *4*, 2883–2887. [CrossRef] [PubMed]

33.   Arca, M.; Aragoni, M.C.; Devillanova, F.A.; Garau, A.; Isaia, F.; Lippolis, V.; Mancini, A.; Verani, G. Structure-Activity Relationships of Synthetic Coumarins as HIV-1 Inhibitors. *Bioinorg. Chem. Appl.* **2006**, *2006*, 1–9. [CrossRef] [PubMed]

34.   Rabie, U.M.; Abou-El-Wafa, M.H.; Nassar, H. In vitro simulation of the chemical scenario of the action of an anti-thyroid drug: Charge transfer interaction of thiazolidine-2-thione with iodine. *Spectrochim. Acta A Mol. Biomol. Spectrosc.* **2011**, *78*, 512–517. [CrossRef] [PubMed]

35.   Refat, M.S.; El-Hawary, W.F.; Moussa, M.A. IR, 1H NMR, mass, XRD and TGA/DTA investigations on the ciprofloxacin/iodine charge-transfer complex. *Spectrochim. Acta A Mol. Biomol. Spectrosc.* **2011**, *78*, 1356–1363. [CrossRef] [PubMed]

36.   El-Sheshtawy, H.S.; Salman, H.M.A.; El-Kemary, M. Halogen vs hydrogen bonding in thiazoline-2-thione stabilization with σ- and π-electron acceptors adducts: Theoretical and experimental study. *Spectrochim. Acta Part A Mol. Biomol. Spectrosc.* **2015**, *137*, 442–449. [CrossRef] [PubMed]

37.   Punyani, S.; Narayana, P.; Singh, H.; Vasudevan, P. Iodine based water disinfection: A review. *J. Sci. Ind. Res. (India)* **2006**, *65*, 116–120.

38. Kulkarni, P.V.; Arora, V.; Rooney, A.S.; White, C.; Bennett, M.; Antich, P.P.; Bonte, F.J. Radiolabeled probes for imaging Alzheimer's plaques. *Nucl. Instrum. Methods Phys. Res. B* **2005**, *B241*, 676–680. [CrossRef]

39. Duan, Y.; Tang, Q.; Chen, Y.; Zhao, Z.; Lv, Y.; Hou, M.; Yang, P.; He, B.; Yu, L. Solid-state dye-sensitized solar cells from poly(ethylene oxide)/polyaniline electrolytes with catalytic and hole-transporting characteristics. *J. Mater. Chem. A* **2015**, *3*, 5368–5374. [CrossRef]

40. Matsumoto, S.; Sumida, R.; Tan, S.E.; Akazome, M. Synthesis of Iodinated Thiazolo[2,3-a]isoquinolinium salts and their Crystal Structures with/without Halogen Bond. *Heterocycles* **2018**, *97*, 755–775. [CrossRef]

41. Kuznetsova, E.A.; Zhuravlev, S.V.; Stepanova, T.N. Synthesis and Properties of Derivatives of 2-Mercaptobenzothiazole. VIII. (Perhydro-1,3-thiazino)[2,3-b]benzothiazolines. *Chem. Heterocycl. Comp.* **1969**, *5*, 467–469. [CrossRef]

42. Gregory, J.; Wei, P.H.L. Thiazolo[2,3-b]benzo(and azobenzo)-thiazole Derivatives, Process for their Preparation and Pharmaceutical Compositions Containing Them. EP Patent 0021773, 1 January 1981.

43. Beard, C.C. 5(6)-Benzene Ring Substituted Benzimidazole-2-carbamate Derivatives Having Anti-protozoal Activity. U.S. Patent 4031228, 22 June 1977.

44. Clark, A.D.; Sykes, P. Thiazolinium salts and their reactions with nucleophiles. *J. Chem. Soc.* **1971**, 103–110. [CrossRef]

45. Deplano, P.; Devillanova, F.; Francesco, A.; Ferraro, J.R.; Mercuri, M.L.; Lippolis, V.; Trogu, E.F. FT-Raman Study on Charge-Transfer Polyiodide Complexes and Comparison with Resonance Raman Results. *Appl. Spectrosc.* **1994**, *48*, 1236–1241. [CrossRef]

46. Bartashevich, E.V.; Yushina, I.D.; Stash, A.I.; Tsirelson, V.G. Halogen Bonding and Other Iodine Interactions in Crystals of Dihydrothiazolo(oxazino)quinolinium Oligoiodides from the Electron-Density Viewpoint. *Cryst. Growth Des.* **2014**, *14*, 5674–5684. [CrossRef]

47. Yu, H.; Yan, L.; He, Y.; Meng, H.; Huang, W. An Unusual Photoconductive Property of Polyiodide and Enhancement by Catenating with 3-Thiophenemethylamine Salt. *Chem. Commun.* **2017**, *53*, 432–435. [CrossRef] [PubMed]

48. Bartashevich, E.V.; Mukhitdinova, S.E.; Yushina, I.D.; Tsirelson, V.G. Electronic criterion for categorizing the chalcogen and halogen bonds: Sulfur–iodine interactions in crystals. *Acta Crystallogr. Sect. B* **2019**, *B75*, 117–126. [CrossRef]

49. Tarasova, N.M.; Kim, D.G. Synthesis and Halocyclization of Allyl Derivatives of 4,5-Dihydro-1,3-thiazol-2-thione. *Bull. South Ural State Univ. Sect. Chem.* **2015**, *7*, 4–10.

50. Bruker. *SMART and SAINT-Plus, Versions 5.0*; Data Collection and Processing Software for the SMART System; Bruker AXS Inc.: Madison, WI, USA, 1998.

51. Bruker. *SHELXTL/PC, Versions 5.10*; An Integrated System for Solving, Refining and Displaying Crystal Structures from Diffraction Data; Bruker AXS Inc.: Madison, WI, USA, 1998.

52. Dolomanov, O.V.; Bourhis, L.J.; Gildea, R.J.; Howard, J.A.K.; Puschmann, H. OLEX2: A Complete Structure Solution, Refinement and Analysis Program. *J. Appl. Cryst.* **2009**, *42*, 339–341. [CrossRef]

53. Yushina, I.D.; Kolesov, B.A.; Bartashevich, E.V. Raman Spectroscopy Study of New Thia- and Oxazinoquinolinium Triodides. *New J. Chem.* **2015**, *39*, 6163–6170. [CrossRef]

54. Dovesi, R.; Erba, A.; Orlando, R.; Zicovich-Wilson, C.M.; Civalleri, B.; Maschio, L.; Rerat, M.; Casassa, S.; Baima, J.; Salustro, S.; et al. Quantum-mechanical condensed matter simulations with CRYSTAL. *Wires Comput. Mol. Sci.* **2018**, *8*, e1360. [CrossRef]

55. Iodine Basis Set. Available online: http://www.tcm.phy.cam.ac.uk/~{}mdt26/basis_sets/I_basis.txt (accessed on 27 September 2019).

56. Gatti, C.; Saunders, V.R.; Roetti, C. Crystal field effects on the topological properties of the electron density in molecular crystals: The case of urea. *J. Chem. Phys.* **1994**, *101*, 10686–10696. [CrossRef]

57. Gatti, C.; Casassa, S. *TOPOND14 User's Manual*; CNR-ISTM of Milano: Milano, Italy, 2013.

58. Clark, T. σ-Holes. *Wires Comput. Mol. Sci.* **2013**, *3*, 13–20. [CrossRef]

59. Bartashevich, E.V.; Yushina, I.D.; Kropotina, K.K.; Muhitdinova, S.E.; Tsirelson, V.G. Testing the Tools for Revealing and Characterizing the Iodine-Iodine Halogen Bond in Crystals. *Acta Crystallogr. Sect. B* **2017**, *B73*, 217–226. [CrossRef] [PubMed]

60. Bol'shakov, O.I.; Yushina, I.D.; Bartashevich, E.V.; Nelyubina, Y.V.; Aysin, R.R.; Rakitin, O.A. Asymmetric triiodide-diiodine interactions in the crystal of (Z)-4-chloro-5-((2-((4-chloro-5H-1,2,3-dithiazol-5-ylidene) amino)phenyl)amino)-1,2,3-dithiazol-1-ium oligoiodide. *Struct. Chem.* **2017**, *28*, 1927–1934. [CrossRef]

61. Tsirelson, V.G.; Avilov, A.S.; Lepeshov, G.G.; Kulygin, A.K.; Stahn, J.; Pietsch, U.; Spence, J.C.H. Quantitative Analysis of the Electrostatic Potential in Rock-Salt Crystals Using Accurate Electron Diffraction Data. *J. Phys. Chem. B* **2001**, *105*, 5068–5074. [CrossRef]

62. Bartashevich, E.V.; Grigoreva, E.A.; Yushina, I.D.; Bulatova, L.M.; Tsirelson, V.G. Modern level for properties prediction of iodine-containing organic compounds: The halogen bonds formed by iodine. *Russ. Chem. Bull.* **2017**, *66*, 1345–1356. [CrossRef]

63. Arca, M.; Aragoni, M.C.; Devillanova, F.A.; Garau, A.; Isaia, F.; Lippolis, V.; Mancini, A.; Verani, G. Reactions between chalcogen donors and dihalogens/interalogens: Typology of products and their characterization by FT-Raman spectroscopy. *Bioinorg. Chem. Appl.* **2006**, *2006*, 58937. [CrossRef] [PubMed]

64. Yushina, I.D.; Rudakov, B.V.; Krivtsov, I.V.; Bartashevich, E.V. Thermal decomposition of tetraalkylammonium iodides. *J. Anal. Calorim.* **2014**, *118*, 425–429. [CrossRef]

65. Yushina, I.D.; Pikhulya, D.G.; Bartashevich, E.V. The features of iodine loss at high temperatures. The case study of crystalline thiazoloquinolinium polyiodides. *J. Therm. Anal. Calorim.* **2019**. [CrossRef]

66. Wang, H.; Liu, J.; Wang, W. Intermolecular and very strong intramolecular C–SeO/N chalcogen bonds in nitrophenyl selenocyanate crystals. *Phys. Chem. Chem. Phys.* **2018**, *20*, 5227–5234. [CrossRef]

Article

Dihydrogen Bonds in Salts of Boron Cluster Anions $[B_nH_n]^{2-}$ with Protonated Heterocyclic Organic Bases

Varvara V. Avdeeva [1,*], Anna V. Vologzhanina [2], Elena A. Malinina [1] and Nikolai T. Kuznetsov [1]

[1] Kurnakov Institute of General and Inorganic Chemistry, Russian Academy of Sciences, Leninskii pr. 31, 119991 Moscow, Russia

[2] Nesmeyanov Institute of Organoelement Compounds, Russian Academy of Sciences, ul. Vavilova 28, 119991 Moscow, Russia

* Correspondence: avdeeva.varvara@mail.ru; Tel.: +7-(495)-954-12-79

Received: 24 May 2019; Accepted: 27 June 2019; Published: 28 June 2019

Abstract: Dihydrogen bonds attract much attention as unconventional hydrogen bonds between strong donors of H-bonding and polyhedral (car)borane cages with delocalized charge density. Salts of *closo*-borate anions $[B_{10}H_{10}]^{2-}$ and $[B_{12}H_{12}]^{2-}$ with protonated organic ligands 2,2'-dipyridylamine (BPA), 1,10-phenanthroline (Phen), and rhodamine 6G (Rh6G) were selectively synthesized to investigate N−H...H−B intermolecular bonding. It was found that the salts contain monoprotonated and/or diprotonated N-containing cations at different ratios. Protonation of the ligands can be implemented in an acidic medium or in water because of hydrolysis of metal cations resulting in the release of H_3O^+ cations into the reaction solution. Six novel compounds were characterized by X-ray diffraction and FT-IR spectroscopy. It was found that strong dihydrogen bonds manifest themselves in FT-IR spectra that allows one to use this technique even in the absence of crystallographic data.

Keywords: boron cages; dihydrogen bonds; hirshfeld surface

1. Introduction

The term "secondary bonds" was introduced by Alcock almost 50 years ago [1] when he described weak bonds found in crystals of inorganic compounds which are shorter than the sum of the wan der Waals radii of the atoms involved in the interaction but longer than the covalent bonds. These bonds are also called non-bonding or non-valent specific bonds [2–5]. Among all types of non-bonding bonds, hydrogen bonds were the only ones to be considered in detail for a long time, as this type of bond is the most important among intermolecular interactions [6–8]. Hydrogen bonds are attractions arising between a hydrogen bond bound with a more electronegative atom (a hydrogen bond donor) and another atom bearing a long pair of electrons (a hydrogen bond acceptor).

There is no doubt that these interactions play a critical role in the organization of the structure of solid compounds and packing, which directly affects their properties. A great number of studies are devoted to this problem, and this topic remains of current interest.

We focused on so-called dihydrogen bonds (DHB) that arise between B–H groups of boron clusters and a protic hydrogen moiety (H–X). Boron clusters (boranes [9–14], carboranes [15], metalloboranes [16,17], and their derivatives) belong to boron hydrides and tend to form numerous DHB with the H–X groups (X = C, O, N) of cations, ligands or solvent molecules found in crystals of their compounds; some of them have been highlighted in a number of studies [18–24] and generalized in reviews [25–28]. The indirect evidence of DHB formed is provided by IR spectroscopy when analyzing the region of the stretching vibrations of the BH groups (2500–2100 cm^{-1}). It should be noted that the perhalogenated boron clusters also participate in non-bonding interactions and can be identified by spectroscopic methods. In particular, numerous B–Cl . . . H–X and B–Cl . . . X interactions were found in salts and complexes of the $[B_{10}Cl_{10}]^{2-}$ anion by analyzing the ^{35}Cl NQR spectroscopy and X-ray diffraction data of the products obtained [29–31].

The most common type of DHB found in compounds of the $[B_nH_n]^{2-}$ boron clusters (n = 10, 12) are the B–H . . . H–N bonds realized between the BH groups of the boron clusters and N-containing cations or molecules. The closest to the present study's detailed discussion of this type of bonds was reported for salts of the *closo*-decaborate anion with monoprotonated and diprotonated bipyridine $[HBipy]_2[B_{10}H_{10}]$ and $[H_2Bipy][B_{10}H_{10}]$ [32,33], respectively, which were studied by X-ray diffraction; atomic charges were determined by the Hirshfeld's method. The authors reported the difference density maps and discussed the electron distribution of the electron density and the electron transfer between the boron cluster and the cation.

Here, we consider the B–H . . . H–N DHB found in salts of the $[B_nH_n]^{2-}$ boron clusters (n = 10, 12) and N-containing monoprotonated and diprotonated organic bases (2,2'-dipyridylamine (BPA), 1,10-phenanthroline (Phen), and rhodamine 6G (Rh6G)); the compounds have been studied by X-ray diffraction techniques and IR spectroscopy.

2. Materials and Methods

Elemental analysis for carbon, hydrogen, and nitrogen content was performed using a CHNS-3 FA 1108 automated elemental analyzer (Carlo Erba Instruments, Milan, Italy). Boron content was determined on an iCAP 6300 Duo ICP emission spectrometer (Thermo Scientific, Waltham, USA) with inductively coupled plasma. The samples were dried in a vacuum at room temperature to a constant weight, thus obtaining solvent-free forms.

IR spectra of the crystals obtained were recorded on an Infralum FT-02 Fourier-transform spectrophotometer (Lumex, St. Petersburg, Russia) in the range of 4000–600 cm^{-1} at a resolution of 1 cm^{-1}. Samples were prepared as Nujol mulls or were recorded in thin layer; NaCl pellets were used.

2.1. Synthesis

DMF (HPLC), acetonitrile (HPLC), CF_3COOH as well as solid BPA, Phen, rhodamine Rh6G·HCl, and $CuSO_4\cdot5\,H_2O$ were purchased from Sigma-Aldrich and used without additional purification. $[Et_3NH]_2[B_{10}H_{10}]$ and $[Et_3NH]_2[B_{12}H_{12}]$ were prepared from decaborane-14 using the known synthetic procedures [34,35].

$(HBPA)_2(H_2BPA)[B_{10}H_{10}]_2$ (1)

A solution of $[Et_3NH]_2[B_{10}H_{10}]$ (0.4 mmol) in water (10 mL) was added to a solution of BPA (0.8 mmol) in acetonitrile (10 mL). Glacial trifluoroacetic acid CF_3COOH (5 mL) was added dropwise to the resulting mixture when stirring. The reaction mixture was allowed to stand at room temperature in a beaker covered with a watch glass. Yellow crystals **1**·$2H_2O$ precipitated after 48 h from the corresponding reaction mixture, which were filtered off, washed with water (2 × 5 mL), and dried in air. Yield, 82%. **1**: $C_{30}H_{51}N_9B_{20}$: Calculated (%): C, 47.79; H, 6.82; N, 16.72; B, 28.7. Found: C, 47.71; H, 6.83; N, 16.69; B, 28.5. IR, cm^{-1} (Nujol mull): 3286s, 3236, 3204, 3136, 2543, 2492, 2444, 2420, 1641s, 1591s, 1557, 1528, 1468s, 1435, 1378s, 1237, 1162, 1108, 1061, 1013, 904, 839, 787s, 781, 773s.

$[PhenH]_2[B_{12}H_{12}]$ (2)

A solution of $[Et_3NH]_2[B_{12}H_{12}]$ (0.4 mmol) in water (10 mL) was added to a solution of Phen (0.8 mmol), respectively, in acetonitrile (10 mL). Glacial trifluoroacetic acid CF_3COOH (5 mL) was added dropwise to the resulting mixtures. The reaction mixture was allowed to stand at room temperature in a beaker covered with a watch glass. Yellow crystals **2**·$2H_2O$ precipitated after 48 h, which were filtered off, washed with acetonitrile (2 × 5 mL), and dried in air. Yield, 77%. **2**: $C_{24}H_{30}N_4B_{12}$: Calculated (%): C, 57.17; H, 6.00; N, 11.11; B, 25.7. Found: C, 57.15; H, 6.02; N, 11.09; B, 25.6. IR, cm^{-1} (Nujol mull): 3098, 3063, 2487, 2443, 1629, 1608, 1586, 1522s, 1461s, 1431s, 1416s, 1226, 1056s, 875, 851s, 783, 738s, 719.

(NHEt$_3$)(HBPA)[B$_{10}$H$_{10}$] (3)

A solution of [Et$_3$NH]$_2$[B$_{10}$H$_{10}$] (0.4 mmol) in acetonitrile (10 mL) was added to a solution of BPA (0.4 mmol) in the same solvent (10 mL). Glacial trifluoroacetic acid CF$_3$COOH (5 mL) was added dropwise when stirring the reaction mixture. The reaction mixture was allowed to stand at room temperature. Yellow crystals 3·2CH$_3$CN started to precipitate after 30 min; after 2 h, they were filtered off, washed with acetonitrile (2 × 5 mL), and dried in air. Yield, 55%. **3**: C$_{16}$H$_{36}$N$_4$B$_{10}$: Calculated (%): C, 48.95; H, 9.24; N, 14.27; B, 27.5. Found: C, 48.96; H, 9.22; N, 14.25; B, 27.2. IR, cm^{-1} (Nujol mull): 3467, 3316, 3255, 3212, 3145, 3111, 2499, 2466, 2448, 2404, 1648s, 1612, 1591s, 1528s, 1488s, 1466s, 1435, 1378, 1273, 1165s, 1018, 773s, 721.

[PhenH)]$_2$[B$_{10}$H$_{10}$] (4)

A solution of [Et$_3$NH]$_2$[B$_{10}$H$_{10}$] (0.4 mmol) in acetonitrile (10 mL) was added to a solution of Phen (0.8 mmol) in the same solvent (10 mL). Glacial trifluoroacetic acid CF$_3$COOH (5 mL) was added dropwise. The reaction mixture was allowed to stand at room temperature in a beaker covered with a watch glass. Yellow solvent-free crystals **4** precipitated after 24 h, which were filtered off, washed with acetonitrile (2 × 5 mL), and dried in air. Yield, 68%. C$_{24}$H$_{28}$N$_4$B$_{10}$: Calculated (%): C, 59.98; H, 5.87; N, 11.66; B, 22.5. Found: C, 59.99; H, 5.81; N, 11.59; B, 22.4. IR, cm^{-1} (thin layer): 3095, 3071, 30545, 3029, 2510, 2477, 2451, 2421, 1658, 1631, 1616s, 1595s, 1540s, 1468s, 1418, 1378s, 1316, 1285, 1214s, 1190, 1159, 1010s, 855, 844, 769, 719s, 715, 665, 622s.

[Rh6GH]$_2$[B$_{12}$H$_{12}$] (5)

A solution of R6G·HCl (0.4 mmol) in water (10 mL) was added to a solution of [Et$_3$NH]$_2$[B$_{12}$H$_{12}$] (0.8 mmol) in acetonitrile (10 mL). The reaction mixture was allowed to stand at room temperature in a beaker covered with a watch glass. A red precipitate was formed after 24 h, which was filtered off, washed with acetonitrile (2 × 5 mL), and dried in air. Red crystals 5·2CH$_3$CN were recrystallized from acetonitrile. Yield, 77%. **5**: C$_{56}$H$_{74}$N$_4$O$_6$B$_{12}$: Calculated (%): C, 65.37; H, 7.25; N, 5.45; B, 12.6. Found: C, 65.41; H, 7.03; N, 5.39; B, 12.5. IR, cm^{-1} (Nujol mull): 3410, 3361, 2484, 2461, 2447, 2422, 2251w, 1727s, 1648, 1607s, 1571, 1549, 1531, 1500, 1467, 1366, 1305, 1255, 1191, 1146, 1079, 1056, 1026, 900, 847, 812, 785, 738.

[PhenH]$_2$(Phen)$_{2.5}$[B$_{10}$H$_{10}$] (6) and [Co(Phen)$_3$][B$_{10}$H$_{10}$] (7)

Compound **6** was obtained in the course of cobalt(II) complexation reaction proceeded in water. We used the procedure similar to those indicated in [36] but the reaction was carried out in water instead of acetonitrile or DMF as follows. A solution of [Et$_3$NH]$_2$[B$_{10}$H$_{10}$] (0.4 mmol) in water (10 mL) was added to a solution containing CoCl$_2$·6 H$_2$O (0.4 mmol) and Phen (1.2 mmol) in water (10 mL). Orange precipitate [Co(Phen)$_3$][B$_{10}$H$_{10}$] (7) was formed in good yield (85%), which was filtered off and dried. After 48 h, light-yellow crystals 6·0.5H$_2$O precipitated from the mother solution; they were filtered off and dried in air. Yield of **6**, < 10%. **6**: IR, cm^{-1} (Nujol mull): 3164, 3152, 3027, 2495, 2476, 2414, 1656, 1651, 1618, 1610, 1543, 824s, 761s. C$_{54}$H$_{48}$N$_9$B$_{10}$: Calculated (%): C, 34.72; H, 2.59; N, 33.75; B, 28.9. Found: C, 34.70; H, 2.57; N, 33.71; B, 28.7. **7**: C$_{36}$H$_{34}$N$_6$B$_{10}$Co: Calculated (%): C, 60.24; H, 4.77; N, 11.71; B, 15.0. Found: C, 60.21; H, 4.71; N, 11, 73; B, 14.9. IR, cm^{-1} (Nujol mull): 3052w; 2469, 2434; 1625, 1582w, 1519s, 1463s, 1427s, 1379, 1342w, 1225w, 1143w, 1103, 861s; 1009; 844s, 727s.

2.2. X-ray Diffraction

Experimental data for 1·2H$_2$O, 2·2H$_2$O, 3·2CH$_3$CN, **4**, 5·CH$_3$CN, and 6·0.5H$_2$O were collected at low temperatures on Bruker Apex II CCD diffractometer using graphite monochromatted MoKα (**1**, **3**, **5**) radiation or CuKα (**2**) radiation with multilayer optics. Intensities of reflections for **4** and **6** were obtained on Bruker Smart Apex CCD diffractometer. The structures were solved by direct method and refined by full-matrix least squares against F^2. Non-hydrogen atoms were refined anisotropically except some disordered carbon and boron atoms. A BPA cation in **3** and a boron cage in solid **6** are disordered over two sites, and were refined isotropically. A number of EADP and ISOR instructions were applied to refine some moieties. Positions of H(C) and H(B) atoms were calculated, and those of H(O) and H(N) atoms were located on difference Fourier maps, and then fixed at 0.87 and 0.88 Å. All hydrogen atoms were included in the refinement by the riding model with $U_{iso}(H) = 1.5U_{eq}(X)$ for methyl groups and water molecules, and $1.2U_{eq}(X)$ for the other atoms. All calculations were made using the SHELXL2014 [37] and OLEX2 [38] program packages. The crystallographic data and experimental details are listed in Table S1 (see Supporting Information).

The crystallographic data for **1**·2H$_2$O, **2**·2H$_2$O, **3**·2CH$_3$CN, **4**, **5**·CH$_3$CN, and **6**·0.5H$_2$O have been deposited with the Cambridge Crystallographic Data Centre as supplementary publications under the CCDC numbers 1917700–1917705. This information may be obtained free of charge from the Cambridge Crystallographic Data Centre via www.ccdc.cam.ac.uk/structures. Hirshfeld surfaces were depicted with CrystalExplorer2.0 [39].

The X-ray powder diffraction patterns of compounds **4**, **5**·2CH$_3$CN, and **7** were measured on a Bruker D8 Advance diffractometer (CuKα1) at RT with LynxEye detector and Ge(111) monochromator, θ/2θ scan from 5° to 80°, step size 0.01125° at the Shared Equipment Center of the Kurnakov Institute. The measurements were performed in transmission mode with the sample deposited in the single-crystal oriented silicon cuvette. The crystals were thoroughly triturated before measurements. The X-ray powder diffraction patterns for **4**, **5**·2CH$_3$CN, and **7** are shown in Figures S11–S13 (see Supplementary Materials).

3. Results and Discussion

3.1. Synthesis and Structures of Salts of the Boron Clusters with Monoprotonated and Diprotonated Organic Bases

The target salts of the boron cluster anions and N-containing cations were synthesized by the reaction between salts [Et$_3$NH]$_2$[B$_{10}$H$_{10}$] or [Et$_3$NH]$_2$[B$_{12}$H$_{12}$] and neutral organic bases Phen, BPA in the presence of CF$_3$COOH in CH$_3$CN and CH$_3$CN/water. The reactions proceeded according to the general reaction scheme:

$$[B_nH_n]^{2-} + 2\,L + CF_3COOH \rightarrow LH^+ + LH_2^+ + [B_nH_n]^{2-}$$

$$(L = Phen, BPA; n = 10, 12)$$

For Rh6G, the reaction proceeded between Rh6G·HCl and [Et$_3$NH]$_2$[B$_{12}$H$_{12}$] in acetonitrile.

It was found that the protonation of the N-containing organic bases L proceeded non-selectively; monoprotonated and diprotonated cations were present in the reaction solution simultaneously and the composition of the final product varied. The majority of crystals of the final products contained the corresponding solvent molecules.

When the reaction proceeded in CH$_3$CN/water in the acidic medium, salts (HBPA)$_2$(H$_2$BPA)[B$_{10}$H$_{10}$]$_2$·2H$_2$O (**1**·2H$_2$O) and [PhenH]$_2$[B$_{12}$H$_{12}$]·2H$_2$O (**2**·2H$_2$O) were isolated which contained diprotonated and/or monoprotonated ligands.

Asymmetric unit of **1**·2H$_2$O contains two monoprotonated cations HBPA$^+$, a diprotonated cation H$_2$BPA^{2+}, two [B$_{10}$H$_{10}$]$^{2-}$ anions and two water molecules (Figure 1a). Cations HBPA$^+$ are flat as their pyridyl fragments take part in intramolecular hydrogen bonding N–H...N (r(N...N) = 2.593(9) and 2.586(9) Å, NHN 133.6° and 133.4°). Both HBPA$^+$ are also connected with water molecules via the N–H...O hydrogen bonds with bridge amino groups (r(N...O) = 2.727(6) and 2.751(6) Å, NHO = 158.7 and 162.0°). All OH groups of water molecules are directed to equatorial atoms of the boron cluster anions (H...H 1.94–2.26, H...B 2.45–2.57 Å). The H$_2$BPA^{2+} cation is non-planar; the dihedral angle between planes of the pyridine cycles is 44.0(3)°. The NH groups of the pyridine cycles form very short contacts with the apical BH groups of the boron cluster anions, additionally supported with a longer bifurcate B–H...H(N)...H–B dihydrogen bonding between a bridging NH-group and equatorial atoms of boron cages.

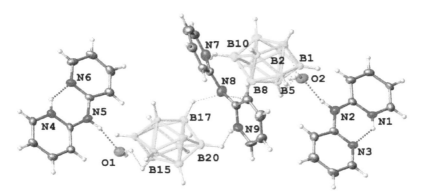

Figure 1. Asymmetric unit of **1**·2H$_2$O in representation of atoms with thermal ellipsoids (p = 50%). H-bonds are depicted with dotted lines.

A similar reaction with the *closo*-dodecaborate anion and Phen resulted in isolation of salt **2**·2H$_2$O, which contained two monoprotonated HPhen$^+$ molecules per one *closo*-dodecaborate anion. Asymmetric unit of this salt contains one cation, two water molecules and half of anion (Figure 2). Water molecules form a dimer through a H-bond (r(O...O) = 2.742(5) Å, OHO = 173.2°). One of two water molecules is involved in H-bonding with two HPhen$^+$ cations (r(N...O) = 2.802(3)–2.125(4) Å, OHN = 120.9–152.6°). The second forms dihydrogen bonds with two anions.

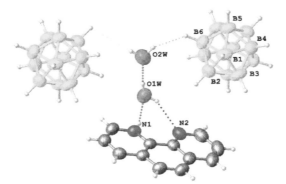

Figure 2. Fragment of the structure of **2**·2H$_2$O in representation of atoms with thermal ellipsoids (p = 50%). Only symmetrically independent boron atoms are labeled. H-bonds are depicted with dotted lines.

As it follows from crystal structures of **1**·2H$_2$O and **2**·2H$_2$O, water molecules both act as likely acceptors of hydrogen bonding towards H(N) atoms to form dihydrogen bonds, and as more likely donor groups towards boron cages as compared with protonated ligands. Thus, a series of salts was obtained in anhydrous reaction conditions. When the reaction under discussion was carried out in the acetonitrile/CF$_3$COOH system, compounds (Et$_3$NH)(HBPA)[B$_{10}$H$_{10}$]·2CH$_3$CN (**3**·2CH$_3$CN) and [PhenH)]$_2$[B$_{10}$H$_{10}$] (**4**) were obtained for BPA and Phen, respectively. No diprotonated cations were found in the compounds synthesized.

Asymmetric unit of **3**·2CH$_3$CN contains two organic cations, HBPA$^+$ and Et$_3$NH$^+$, the *closo*-decaborate anion and two acetonitrile molecules (Figure 3). Both cations were found to be disordered, and the disorder could not be resolved in non-centrosymmetric groups of lower symmetry. The distorted geometry and high errors do not allow us to unambiguously determine positions of H(N) atoms in the amino-pyridilium cation, but planar conformation of HBPA$^+$ allows us to propose intramolecular hydrogen bonding N–H...N, while the amino groups N(3)–H and N(1)–H form dihydrogen N–H...H–B bonds with equatorial atoms of the *closo*-decaborate anion. The Et$_3$NH$^+$ cation interacts with apical atoms of the anion. Acetonitrile molecules are involved in weak C–H...N bonding with methyl groups of Et$_3$NH$^+$.

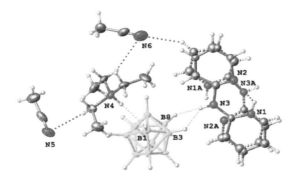

Figure 3. Asymmetric unit of **3**·2CH$_3$CN in representation of atoms with thermal ellipsoids (p = 50%). H-bonds are depicted with dotted lines.

When Phen was allowed to react with the *closo*-decaborate anion in acetonitrile in the presence of trifluoroacetic acid, a solvent-free salt [PhenH)]$_2$[B$_{10}$H$_{10}$] (**4**) was isolated. This compound contains two independent [B$_{10}$H$_{10}$]$^{2-}$ anions and four PhenH$^+$ cations (Figure 4). The independent *closo*-decaborate anions have different environments. The B(1)–B(10) anion is situated in a hydrophobic 'cavity' formed by cations; and takes part in B–H . . . H–C and B–H . . . π interactions only. The B(11)–B(20) anion besides these hydrophobic interactions forms also contacts with the NH-groups via equatorial and apical boron atoms with three cations. The intramolecular N–H...N hydrogen bond is found in all cations (r(N...N) = 2.697(6)–2.723(6) Å, NHN = 104°–105°). In addition, the NH groups of the three cations form shortened contacts with the *closo*-borate anions (see above), and one cation does not participate in the contacts.

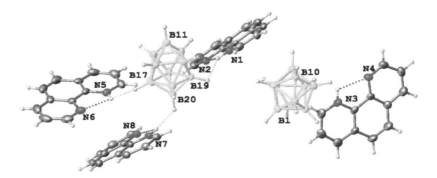

Figure 4. Asymmetric unit of **4** in representation of atoms with thermal ellipsoids (p = 50%). H-bonds are depicted with dotted lines.

For rhodamine Rh6G, we obtained a salt with the [Rh6GH]$_2$[B$_{12}$H$_{12}$]·2CH$_3$CN composition (**5**·2CH$_3$CN). This compound was synthesized in an acetonitrile/water solution when Rh6G·HCl was used and the precipitate obtained was recrystallized from acetonitrile. Asymmetric unit of this salt contains a protonated cation, half the *closo*-dodecaborate anion, and an acetonitrile molecule (Figure 5). The angle between mean planes of the phenyl ring and condensed rings of the Rh6GH$^+$ is 116.9(1)°. Cations and anions from infinite chains connected through dihydrogen N−H...H−B bonds. There exist π...π stacking between parallel Rh6GH$^+$ molecules situated 3.490(1) Å from each other, additionally supported with C=O...H−C interactions.

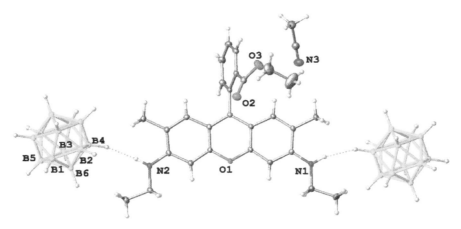

Figure 5. Fragment of the structure of **5**·2CH$_3$CN in representation of atoms with thermal ellipsoids (p = 50%). Only symmetrically independent atoms are labelled. H-bonds are depicted with dotted lines.

Note that acidic media can be created not only by the presence of an acid in reaction mixtures but also because of hydrolysis of metal cations in aqueous solutions. In particular, when we studied the cobalt(II) complexation with Phen in the presence of the *closo*-decaborate anion, two products were isolated, namely *tris*-chelate cobalt(II) complex [Co(phen)$_3$][B$_{10}$H$_{10}$] (**7**) (main product) [36] and a salt of the *closo*-decaborate anion with Phen (**6**). The reaction proceeded in the water/acetonitrile system according to the scheme:

CoCl$_2$ + 3 Phen + [B$_{10}$H$_{10}$]$^{2-}$ → [Co(Phen)$_3$][B$_{10}$H$_{10}$] (**7**) + [H$_2$(Phen)$_{4.5}$][B$_{10}$H$_{10}$] (**6**)

Cobalt(II) complex **7** was characterized by elemental analysis, IR spectroscopy, and powder X-ray diffraction data. It was found that its powder X-ray diffraction pattern coincided with that calculated from the X-ray diffraction data reported [36].

After precipitation of complex **7**, single crystals **6**·0.5H$_2$O were obtained from the reaction mixture in low yield. It was found that they contain salt of the *closo*-decaborate anion with non-protonated and monoprotonated Phen molecules [H$_2$(Phen)$_{4.5}$][B$_{10}$H$_{10}$]. Their preparation is explained by partial hydrolysis of cobalt(II) cation in water according to the following scheme:

[Co(H$_2$O)$_6$]$^{2+}$ + H$_2$O ↔ [Co(OH)(H$_2$O)$_5$]$^+$ + H$_3$O$^+$

[B$_{10}$H$_{10}$]$^-$ + Phen + H$_3$O$^+$ → [H$_2$(Phen)$_{4.5}$][B$_{10}$H$_{10}$] (**6**)

As the concentration of protons in the resulting aqueous reaction solution was significantly lower than that in the case when we used trifluoroacetic acid, final salt **6** contained not only protonated ligand LH$^+$ but neutral ligand L as well.

The asymmetric unit of compound **6**·0.5H$_2$O contains four and a half independent Phen molecules, a disordered *closo*-decaborate anion, and partially occupied position of water molecule (Figure 6a). None of H(N) atoms could be located on difference Fourier maps due to low crystal quality, and at the same time their positions can be proposed from the most likely intermolecular hydrogen bonds.

Three of Phen molecules are packed in stacks with an interplanar distance of ~3.5 Å, and these stacks are connected by means of N−H...N bonds (r(N...N) = 2.82(2) Å, NHN = 159.4°, only of two H-bonded stacks should be charged in this case, thus, positions of H(N) protons are only half occupied, Figure 6b). Besides, there exists a Phen molecule connected with a water molecule by means of H-bonds (r(N...O) = 2.55(2) Å, NHO = 168.9°), while hydrogen atoms of the water molecule should be directed towards boron atoms of two cages. Thus, the only Phen molecule surrounded by H(C) and H(B) atoms is neutral.

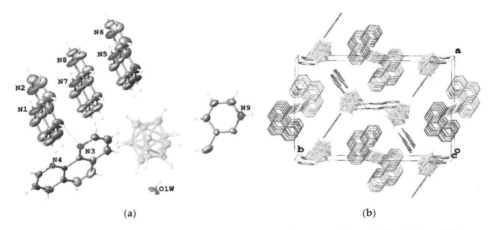

(a)　　　　　　　　　　　　　　　(b)

Figure 6. (a) Asymmetric unit of **6** · 0.5H$_2$O in representation of atoms with thermal ellipsoids (p = 50%). (b) Fragment of unit cell of **6** · 0.5H$_2$O along crystallographic axis c. H(C) atoms are omitted, N-H...N bonds are dashed.

Therefore, when studying complexation of metals in water in the presence of ligands capable to be protonated, it should be taken into account that the corresponding salts can be obtained with anions present in reaction mixtures rather than metal complexes because metals can be reversible hydrolyzed in aqueous solutions.

3.2. Hirshfeld Surface Analysis

Molecular Hirshfeld surface analysis is a convenient tool to rank the week intermolecular interactions. The surfaces were previously used to investigate dihydrogen bonding in borane salts [40,41] and metallocarboranes [42,43]. For the compounds studied, dihydrogen bonds form 86–99% of molecular surface. At the same time, both the nature of donor groups, and that of a boron cage manifest themselves on Hirshfeld surfaces mapped with d$_{norm}$ (Figure 7). The closest and strongest of intermolecular interactions are depicted in red. In *closo*-decaboranes such interactions appear between the anion and cations or water molecules, which also supports our conclusion that water molecules are likely donors of H-bonding. All B−H...H−O interactions also satisfy criteria of dihydrogen bonding given in Reference [24], particularly, r(H...H) < 2.2 Å and OHH > BHH. B−H...H−N interactions in **1**·2H$_2$O, **3**·2CH$_3$CN and **4** also manifest themselves as red regions on highly curved Hirshfeld surface mapped with d$_{norm}$. In *closo*-dodecaborates, the area of red regions seems to be smaller than that for the same types of interactions in *closo*-decaborates and is more similar to the Hirshfeld surface of metallocarboranes mapped with d$_{norm}$ [43]. Besides B−H...H−N bonds, some B−H...H−C interactions are also among the closest in crystals of **2**·2H$_2$O and **5**·2CH$_3$CN which was characteristic of borane and bisborane salts [40,41].

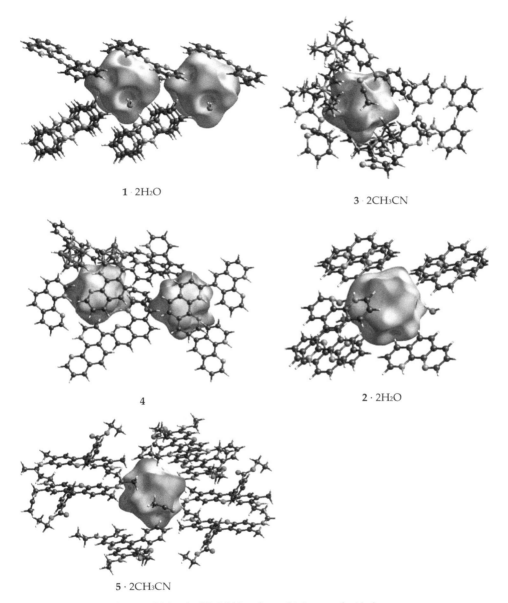

1 · 2H₂O

3 · 2CH₃CN

4

2 · 2H₂O

5 · 2CH₃CN

Figure 7. Molecular Hirshfeld surfaces of **1–5** mapped with d$_{norm}$.

Note that other types of dihydrogen bound could be found in boron-containing compounds. In particular, the M···H–B interactions (M = Li, K) and homopolar CH···HC and BH···HB dihydrogen bonds were found to exist in hydrogen storage materials LiN(CH$_3$)$_2$BH$_3$ and KN(CH$_3$)$_2$BH$_3$ [44]; they were described by the charge and energy decomposition method and the Interacting Quantum Atoms approach. The BH···HB bonds were discussed by the authors as destabilizing, whereas the CH···HC bonds should be considered stabilizing.

3.3. IR Spectroscopy Data of Compounds Synthesized

IR spectroscopy is a perfect tool to determine the presence of dihydrogen bonds in compounds with the boron cluster anions. For compounds containing neutral organic ligands, IR spectroscopy allows one to determine whether an organic ligand is present in a free form, is coordinated by metal atoms or is protonated. IR spectra of salts containing protonated N-containing organic ligands have pronounced changes in regions of characteristic bands of the *closo*-borate anions and ligands L, particularly in both B–H and N–H stretching vibrations, as compared to compounds with free ligands and alkaline *closo*-borates. IR spectra of all compounds discussed are present in SI; selected IR data are shown in Table 1.

Table 1. Maxima of ν(NH) and ν(BH) absorption bands in IR spectra of compounds **1–6** as compared to alkali metal *closo*-borates (cm^{-1}).

Compound	ν(NH)	ν(BH)
$K_2[B_{10}H_{10}]$	—	2529, 2461
Phen	—	—
4	3095	2510, 2477, 2451, 2421
6·0.5H_2O	3164, 3152	2495, 2476, 2414
BPA	3255, 3181, 3102	—
1·2H_2O	3286, 3236, 3204, 3136	2543, 2492, 2444, 2420
3·2CH_3CN	3467, 3316, 3255, 3212, 3145, 3111	2499, 2466, 2448, 2404
$Cs_2[B_{12}H_{12}]$	—	2465
5·CH_3CN	3410, 3361	2484, 2461, 2447, 2422
2·2H_2O	3098, 3063	2487, 2443

In salts with cations [PhenH]$^+$, a new band is observed in the ν(NH) region (3350–3100 cm^{-1}), which is an unambiguous evidence for the protonation of Phen. This band is observed for compound **6** with protons not localized by X-ray diffraction, indicating that NH groups are present in the structure.

Unlike Phen, BPA ligand in the free, non-protonated state contains NH groups which are manifested as three ν(NH) bands. Salts with monoprotonated and diprotonated BPA contain four (for **1**) or five (**3**·2CH_3CN) bands in the IR spectra, respectively, which reflects DHB between the BH groups of the boron cluster anions and NH groups of the ligand. In addition, a ν(NH) band with a maximum at 3467 cm^{-1} is observed in the spectrum of compound **3**·2CH_3CN which is assigned to the triethylammonium cation from the starting salt.

Pronounced changes are observed in IR spectra of the final compounds in the ν(BH) region (2550–2400 cm^{-1}). Boron cluster anions are known to form numerous X–H ... H–B interactions in complexes and salts with cations and neutral molecules containing NH, CH, OH groups [28]. This leads to splitting of the ν(BH) band to several components indicating the violation of the initial state of the BH bonds in "free" (non-coordinated) boron clusters (such as alkaline *closo*-borates).

In the IR spectrum of $K_2[B_{10}H_{10}]$, two bands are observed which correspond to ν(BH) stretching vibrations of apical and equatorial BH groups (Figure 8, curve *1*). In the IR spectra of salts with the $[B_{10}H_{10}]^{2-}$ anion and protonated ligands (curves *2–5*), there is a pronounced splitting of this band, therefore no bands corresponding to stretching vibrations of apical and equatorial BH groups can be defined. In the spectrum of salts with the $[B_{12}H_{12}]^{2-}$ anion (curves *7* and *8*), the ν(BH) band is also split to several components because of DHB, while in the spectrum of $Cs_2[B_{12}H_{12}]$ (curve *6*), one band is observed indicating equivalence of all BH bonds in the B_{12} icosahedron.

In the IR spectra of salts under discussion there are changes in the intensities and redistribution of bands in regions corresponding to vibrations of the phenyl rings of the heterocycles (1700–600 cm^{-1}), namely in the ν(CC) and ν(CN) regions (1650–1500 cm^{-1}) as well as out-of-plane π(CH) vibrations (850–700 cm^{-1}) (see Table S2).

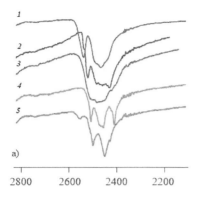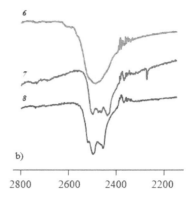

Figure 8. IR spectra (in the ν(BH) region) of compounds with (**a**) $[B_{10}H_{10}]^{2-}$ and (**b**) $[B_{12}H_{12}]^{2-}$ anions in comparison with alkali *closo*-borates: (*1*) $K_2[B_{10}H_{10}]$; (*2*) $[PhenH]_2[B_{10}H_{10}]$ (**4**); (*3*) $[H_2(Phen)_{4.5}][B_{10}H_{10}]$ (**6**); (*4*) $(Et_3NH)(BPAH)[B_{10}H_{10}]$ (**3**); (*5*) $(BPAH)_2(BPAH_2)[B_{10}H_{10}]_2$ (**1**); (*6*) $Cs_2[B_{12}H_{12}]$; (*7*) $[Rh6GH]_2[B_{12}H_{12}]$ (**5**); (*8*) $[PhenH]_2[B_{12}H_{12}]$ (**2**).

4. Conclusions

Six novel salts of the *closo*-decaborate and *closo*-dodecaborate anions with cations of 2,2′-dipyridylamine, 1,10-phenanthroline, and rhodamine 6G were synthesized and studied by IR spectroscopy and X-ray diffraction. Compounds contain monoprotonated and/or diprotonated organic cations depending on the synthetic method. Numerous dihydrogen B–H...H–O and B–H . . . H–N interactions between the boron clusters, water molecules and N-containing cations were detected. Characteristic changes were found in the IR spectra of the compounds indicating unambiguously the formation of dihydrogen bonds.

Supplementary Materials: The following are available online at http://www.mdpi.com/2073-4352/9/7/330/s1, Table S1: Crystallographic data and refinement parameters for crystals **1–6**, Table S2: Maxima of selected absorption bands in IR spectra of compounds **1–6** as compared to alkali metal *closo*-borates, Figure S1: IR spectrum of $K_2[B_{10}H_{10}]\cdot nH_2O$, Figure S2: IR spectrum of bpa, Figure S3: IR spectrum of $(Hbpa)_2(H_2bpa)[B_{10}H_{10}]_2$ (**1**), Figure S4: IR spectrum of $(NHEt_3)(Hbpa)[B_{10}H_{10}]$ (**3**), Figure S5: IR spectrum of phen, Figure S6: IR spectrum of $[Hphen)]_2[B_{10}H_{10}]$ (**4**), Figure S7: IR spectrum of $[H_2(phen)_{4.5}][B_{10}H_{10}]$ (**6**), Figure S8: IR spectrum of $Cs_2[B_{12}H_{12}]$, Figure S9: IR spectrum of $[PhenH]_2[B_{12}H_{12}]$ (**2**), Figure S10: IR spectrum of $[Rh6GH]_2[B_{12}H_{12}]\cdot CH_3CN$ (**5**), Figure S11: Experimental and calculated X-ray diffraction patterns of $[PhenH)]_2[B_{10}H_{10}]$ (**4**); Figure S12: Experimental and calculated X-ray diffraction patterns of $[Rh6GH]_2[B_{12}H_{12}]\cdot CH_3CN$ (**5**); Figure S13: Experimental and calculated X-ray diffraction patterns of complex **7**.

Author Contributions: Synthesis, writing—original draft preparation, review and editing – V.V.A and E.A.M; X-ray diffraction studies and Hirshfeld surface analysis – A.V.V.; conceptualization, supervision – N.T.K.

Funding: "This research received no external funding".

Acknowledgments: This work was performed within the framework of the State Assignment of the Kurnakov Institute (IGIC RAS) in the field of fundamental scientific research. The X-ray diffraction studies were performed within the framework of the State Assignment of the Nesmeyanov Institute (INEOS RAS). X-ray diffraction studies were performed at the Centre for Molecular Studies of the Nesmeyanov Institute; X-ray powder diffraction studies were performed at the Shared Facility Centre of the Kurnakov Institute.

Conflicts of Interest: The authors declare no conflict of interest.

References

1. Alcock, N.W. Secondary Bonding to Nonmetallic Elements. In *Advances in Inorganic Chemistry and Radiochemistry*; Eméleus, H.J., Sharpe, A.G., Eds.; Elsevier: Amsterdam, Netherlands, 1972; Volume 15, pp. 1–444.

2. Glidewell, C. Some chemical and structural consequences of non-bonded interactions. *Inorg. Chim. Acta* **1975**, *12*, 219–227. [CrossRef]

3. Kuz'mina, L.G.; Struchkov, Y.T. Structural chemistry of organomercury compounds – role of secondary interactions. *Croat. Chim. Acta* **1984**, *57*, 701–724.

4. Kuz'mina, L.G. Secondary bonds and their role in chemistry. *Russ. J. Coord. Chem.* **1999**, *25*, 599–617.

5. Virovets, A.V.; Podberezskaya, N.V. Specific nonbonded interactions in the structures of $M_3X_7^{4+}$ and $M_3X_4^{4+}$ (M = Mo, W; X = O, S, Se) clusters. *J. Struct. Chem.* **1993**, *34*, 306–322. [CrossRef]

6. Steiner, T. The Hydrogen Bond in the Solid State. *Angew. Chem. Int. Ed.* **2002**, *41*, 48–76. [CrossRef]

7. George, A.J. *An Introduction to Hydrogen Bonding (Topics in Physical Chemistry)*; Oxford University Press: New York, NY, USA, 1997; pp. 4–320.

8. Muetterties, E.L.; Knoth, W.H. *Polyhedral Boranes*; Marcel Dekker: New York, NY, USA, 1968; pp. 104–168.

9. Greenwood, N.N.; Earnshaw, A. *Chemistry of the Elements*, 2nd ed.; Butterworth–Heinemann: Oxford, UK, 1997; pp. 4–1600.

10. Malinina, E.A.; Avdeeva, V.V.; Goeva, L.V.; Kuznetsov, N.T. Coordination compounds of electron-deficient boron cluster anions $B_nH_n^{2-}$ (n = 6, 10, 12). *Russ. J. Inorg. Chem.* **2010**, *55*, 2148–2202. [CrossRef]

11. Avdeeva, V.V.; Malinina, E.A.; Sivaev, I.B.; Bregadze, V.I.; Kuznetsov, N.T. Silver and Copper Complexes with *closo*-Polyhedral Borane, Carborane and Metallacarborane Anions: Synthesis and X-ray Structure. *Crystals* **2016**, *6*, 60. [CrossRef]

12. Sivaev, I.B.; Prikaznov, A.V.; Naoufal, D. Fifty years of the *closo*-decaborate anion chemistry. *Collect. Czech. Chem. Commun.* **2010**, *75*, 1149–1199. [CrossRef]

13. Sivaev, I.B.; Bregadze, V.I.; Sjoberg, S. Chemistry of *closo*-Dodecaborate Anion $[B_{12}H_{12}]^{2-}$: A Review. *Collect. Czech. Chem. Commun.* **2002**, *67*, 679–727. [CrossRef]

14. Zhizhin, K.Y.; Zhdanov, A.P.; Kuznetsov, N.T. Derivatives of *closo*-decaborate anion $[B_{10}H_{10}]^{2-}$ with *exo*-polyhedral substituents. *Russ. J. Inorg. Chem.* **2010**, *55*, 2089–2127. [CrossRef]

15. Grimes, R. *Carboranes*, 2nd ed.; Academic Press: Cambridge, MA, USA, 2011; 1139p.

16. Klanberg, F.; Muetterties, E.L.; Guggenberger, L.J. Metalloboranes. I. Metal complexes of B3, B9, B9S, B10, and B11 borane anions. *Inorg. Chem.* **1968**, *7*, 2272–2278. [CrossRef]

17. Housecroft, C.E.; Fehlner, T.P. Metalloboranes: Their Relationships to Metal-Hydrocarbon Complexes and Clusters. *Adv. Organomet. Chem.* **1982**, *21*, 57–112.

18. Shubina, E.S.; Bakhmutova, E.V.; Filin, A.M.; Sivaev, I.B.; Teplitskaya, L.N.; Chistyakov, A.L.; Stankevich, I.V.; Bakhmutov, V.I.; Bregadze, V.I.; Epstein, L.M. Dihydrogen bonding of decahydro-*closo*-decaborate(2-) and dodecahydro-*closo*-dodecaborate(2-) anions with proton donors: Experimental and theoretical investigation. *J. Organomet. Chem.* **2002**, *657*, 155–162. [CrossRef]

19. Chen, X.; Zhao, J.-C.; Shore, S.G. The Roles of Dihydrogen Bonds in Amine Borane Chemistry. *Acc. Chem. Res.* **2013**, *46*, 2666–2675. [CrossRef] [PubMed]

20. Mebs, S.; Kalinowski, R.; Grabowsky, S.; Förster, D.; Kickbusch, R.; Justus, E.; Morgenroth, W.; Paulmann, C.; Luger, P.; Gabel, D.; et al. Charge transfer via the dative N-B bond and dihydrogen contacts. Experimental and theoretical electron density studies of four deltahedral boranes. *J. Phys. Chem. A.* **2011**, *115*, 1385–1395. [CrossRef] [PubMed]

21. Crabtree, R.H.; Siegbahn, P.E.M.; Eisenstein, O.; Rheingold, A.L.; Koetzle, T.F. A new intermolecular interaction: unconventional hydrogen bonds with element-hydride bonds as proton acceptor. *Acc. Chem. Res.* **1996**, *29*, 348–354. [CrossRef]

22. Richardson, T.B.; de Gala, S.; Siegbahn, P.E.M.; Crabtree, R.H. Unconventional Hydrogen Bonds: Intermolecular B-H...H-N Interactions. *J. Am. Chem. Soc.* **1995**, *117*, 12875–12876. [CrossRef]

23. Hawthorne, M.F.; Beno, C.L.; Harwell, D.E.; Jalisatgi, S.S.; Knobler, C.B. Intra- and inter-molecular hydrogen bonding in some cobaltacarboranes. *J. Mol. Struct. (Theochem)* **2003**, *656*, 239–247. [CrossRef]

24. Virovets, A.V.; Vakulenko, N.N.; Volkov, V.V.; Podberezskaya, N.V. Crystal structure of di(1-*n*-dodecylpyridine) decahydro-*closo*-decaborate(2-) $(C_5H_5N-C_{12}H_{25})_2[B_{10}H_{10}]$. *J. Struct. Chem* **1994**, *35*, 339–344. [CrossRef]

25. Custelcean, R.; Jackson, J.E. Dihydrogen Bonding: Structures, Energetics, and Dynamics. *Chem. Rev.* **2001**, *101*, 1963–1980. [CrossRef]

26. Belkova, N.V.; Epstein, L.M.; Filippov, O.A.; Shubina, E.S. Hydrogen and dihydrogen bonds in the reactions of metal hydrides. *Chem. Rev.* **2016**, *116*, 8545–8587. [CrossRef] [PubMed]

27. Bakhmutov, V.I. *Dihydrogen Bonds: Principles, Experiments and Applications*; Wiley-Interscience: Hoboken, NJ, USA, 2008.
28. Malinina, E.A.; Avdeeva, V.V.; Goeva, L.V.; Polyakova, I.N.; Kuznetsov, N.T. Specific interactions in metal salts and complexes with cluster boron anions $B_nH_n^{2-}$ ($n = 6, 10, 12$). *Russ. J. Inorg. Chem.* **2011**, *56*, 687–697. [CrossRef]
29. Avdeeva, V.V.; Kravchenko, E.A.; Gippius, A.A.; Vologzhanina, A.V.; Ugolkova, E.A.; Minin, V.V.; Malinina, E.A.; Kuznetsov, N.T. Synthesis, structure, and physicochemical properties of triply-bridged binuclear copper(II) complex $[Cu_2Phen_2(\mu\text{-}CH_3CO_2)_2(\mu\text{-}OH)]_2[B_{10}Cl_{10}]$. *Inorg. Chim. Acta* **2019**, *487*, 208–213. [CrossRef]
30. Kravchenko, E.A.; Gippius, A.A.; Vologzhanina, A.V.; Avdeeva, V.V.; Malinina, E.A.; Ulitin, E.O.; Kuznetsov, N.T. Secondary interactions in decachloro-*closo*-decaborates of alkali metals $M_2[B_{10}Cl_{10}]$ ($M = K^+$ and Cs^+): ^{35}Cl NQR and X-ray studies. *Polyhedron* **2016**, *117*, 561–568. [CrossRef]
31. Kravchenko, E.A.; Gippius, A.A.; Vologzhanina, A.V.; Avdeeva, V.V.; Malinina, E.A.; Demikhov, E.I.; Kuznetsov, N.T. Secondary interactions as defined by ^{35}Cl NQR spectra in cesium decachloro-*closo*-decaborates prepared in non-aqueous solutions. *Polyhedron* **2017**, *138*, 140–144. [CrossRef]
32. Chantler, C.T.; Maslen, E.N. Charge transfer and three-centre bonding in monoprotonated and diprotonated 2,2'-bipyridylium decahydro-*closo*-decaborate(2-). *Acta Crystallogr., Sect. B* **1989**, *45*, 290–297. [CrossRef]
33. Fuller, D.J.; Kepert, D.L.; Skelton, B.W.; White, A.H. Structure, Stereochemistry and Novel 'Hydrogen Bonding' in Two Bipyridinium Salts of the $B_{10}H_{10}^{2-}$ Anion. *Aust. J. Chem.* **1987**, *40*, 2097–2105. [CrossRef]
34. Miller, H.C.; Miller, N.E.; Muetterties, E.L. Synthesis of polyhedral boranes. *J. Am. Chem. Soc.* **1963**, *85*, 3885–3886. [CrossRef]
35. Greenwood, N.N.; Morris, J.H. Novel Synthesis of the $B_{12}H_{12}^{2-}$ Anion. *Proc. Chem. Soc.* **1963**, *11*, 338.
36. Avdeeva, V.V.; Vologzhanina, A.V.; Goeva, L.V.; Malinina, E.A.; Kuznetsov, N.T. Reactivity of boron cluster anions $[B_{10}H_{10}]^{2-}$, $[B_{10}Cl_{10}]^{2-}$ and $[B_{12}H_{12}]^{2-}$ in cobalt(II)/cobalt(III) complexation with 1,10-phenanthroline. *Inorg. Chim. Acta* **2015**, *428*, 154–162. [CrossRef]
37. Sheldrick, G.M. SHELXT—Integrated space group and crystal-structure determination. *Acta Cryst.* **2015**, *A71*, 3–8. [CrossRef] [PubMed]
38. Dolomanov, O.V.; Bourhis, L.J.; Gildea, R.J.; Howard, J.A.K.; Puschmann, H. OLEX2: A complete structure solution, refinement and analysis program. *J. Appl. Cryst.* **2009**, *42*, 339–341.
39. Turner, M.J.; McKinnon, J.J.; Wolff, S.K.; Grimwood, D.J.; Spackman, P.R.; Jayatilaka, D.; Spackman, M.A. CrystalExplorer17 (2017). University of Western Australia. Available online: http://hirshfeldsurface.net (accessed on 27 June 2019).
40. Qi, G.; Wang, K.; Yang, K.; Zou, B. Pressure-Induced Phase Transition of Hydrogen Storage Material Hydrazine Bisborane: Evolution of Dihydrogen Bonds. *J. Phys. Chem. C* **2016**, *120*, 21293–21298. [CrossRef]
41. Qi, G.; Wang, K.; Xiao, G.; Zou, B. High pressure, a protocol to identify the weak dihydrogen bonds: experimental evidence of C–H···H–B interaction. *Sci. Chin. Chem.* **2018**, *61*, 276–280. [CrossRef]
42. Smol'yakov, A.F.; Korlyukov, A.A.; Dolgushin, F.M.; Balagurova, E.V.; Chizhevsky, I.T.; Vologzhanina, A.V. Studies of Multicenter and Intermolecular Dihydrogen B–H···H–C Bonding in $[4,8,8'\text{-}exo\text{-}\{PPh_3Cu\}\text{-}4,8,8'\text{-}(\mu\text{-}H)_3\text{-}commo\text{-}3,3'\text{-}Co(1,2\text{-}C_2B_9H_9)(1',2'\text{-}C_2B_9H_{10})]$. *Eur. J. Inorg. Chem.* **2015**, 5847–5855. [CrossRef]
43. Bennour, I.; Haukka, M.; Teixidor, F.; Vinas, C.; Kabadou, A. Crystal structure and Hirshfeld surface analysis of $[N(CH_3)_4][2,2'\text{-}Fe(1,7\text{-}closo\text{-}C_2B_9H_{11})_2]$. *J. Organomet. Chem.* **2017**, *846*, 74–80. [CrossRef]
44. Sagan, F.; Filas, R.; Mitoraj, M.P. Non-Covalent Interactions in Hydrogen Storage Materials $LiN(CH_3)_2BH_3$ and $KN(CH_3)_2BH_3$. *Crystals* **2016**, *6*. [CrossRef]

Article

Molecular Structures Polymorphism the Role of F . . . F Interactions in Crystal Packing of Fluorinated Tosylates

Dmitry E. Arkhipov [1], Alexander V. Lyubeshkin [2], Alexander D. Volodin [1] and Alexander A. Korlyukov [1,*]

[1] A.N.Nesmeyanov Institute of Organoelement Compounds of Russian Academy of Sciences, 119991 Moscow, Russia; d.arkhipov1988@gmail.com (D.E.A.); alex.d.volodin@gmail.com (A.D.V.)

[2] Federal Scientific Research Center "Crystallography and Photonics" of Russian Academy of Sciences, 119333 Moscow, Russia; cito2006@rambler.ru

* Correspondence: alex@xrlab.ineos.ac; Tel. +499-135-9214

Received: 19 April 2019; Accepted: 5 May 2019; Published: 7 May 2019

Abstract: The peculiarities of interatomic interactions formed by fluorine atoms were studied in four tosylate derivatives $p\text{-}CH_3C_6H_4OSO_2CH_2CF_2CF_3$ and $p\text{-}CH_3C_6H_4OSO_2CH_2(CF_2)_nCHF_2$ (n = 1, 5, 7) using X-ray diffraction and quantum chemical calculations. Compounds $p\text{-}CH_3C_6H_4OSO_2CH_2(CF_2)_nCHF_2$ (n = 1, 5) were crystallized in several polymorph modifications. Analysis of intermolecular bonding was carried out using QTAIM approach and energy partitioning. All compounds are characterized by crystal packing of similar type and the contribution of intermolecular interactions formed by fluorine atoms to lattice energy is raised along with the increase of their amount. The energy of intra- and intermolecular F . . . F interactions is varied in range 0.5–13.0 kJ/mol. Total contribution of F . . . F interactions to lattice energy does not exceed 40%. Crystal structures of studied compounds are stabilized mainly by C-H . . . O and C-H . . . F weak hydrogen bonds. The analysis of intermolecular interactions and lattice energies in polymorphs of $p\text{-}CH_3C_6H_4OSO_2CH_2(CF_2)_nCHF_2$ (n = 1, 5) has shown that most stabilized are characterized by the least contribution of F . . . F interactions.

Keywords: organofluorine compounds; polymorphism; QTAIM; NCI; quantum chemical calculations; lattice energy; intermolecular interactions; F . . . F interactions

1. Introduction

Organosulfur compounds containing fluorinated hydrocarbon moieties are usually considered as dangerous hydrocarbon pollutants that destroy cell membranes [1,2]. Among these compounds, perfluoroalkyl sulfonates, perfluoroalkyl sulfoacids, and sulfamides are the most dangerous and pervasive in environment owing to their high stability and surfactant properties. On the other hand, there are examples of application of above compounds in medicine as drug delivery vehicles [3,4] and antimicrobial agents [5]. The presence of perfluorinated hydrocarbon moiety plays special role in binding of these compounds with biomolecules in solution and in complexes with proteins via hydrophobic interactions. For instance, perfluorooctane sulfuric acid can occupy the position between peptide chains of serum albumin [6] mostly via weak van-der-Waals H . . . F interactions. Besides, perfluoroalkyl chains can form aggregates (molecular ensembles, micelles [7], liquid crystal phases [8]) in which the role of F . . . F interactions can be considerable. The crystal structure can be considered as a model for molecular associations of such compounds; and XRD techniques allow studying the nature of weak intermolecular interactions in detail. The nature of interactions formed by fluorine atoms in organic crystals was extensively studied in many articles [9–15]. Typically, intermolecular interactions in solids containing CF_3 groups or fluorinated aromatic fragments are studied. Only few papers devoted to investigation of compounds with alkyl perfluorinated substituents were published to date [16,17]. Unfortunately, the computational studies in these articles are limited to dimers extracted from crystal packing. In present paper we studied the nature of molecular association in four tosylate derivatives with CF_3 (**1**) and $(CF_2)_nCHF_2$ (n = 1, 5, 7 in **2–4**) groups using single crystal X-ray diffraction and quantum chemical calculations utilizing periodic boundary conditions. Compounds **2** and **3** were crystallized in two polymorphic modifications (**2a**, **2b**, **3a**, and **3b**).

The studied compounds were synthesized as precursors for preparation of fluorinated azides. In turn, they can be used for modification of biologically active molecules such as various antibacterial agents. The compounds described herein contain fluorinated hydrocarbon moiety of different size from short (C_2F_5 and CF_2CHF_2) to long ($(CF_2)_7CHF_2$), thus giving the opportunity to discover and compare the effect of substituent on crystal packing and physicochemical properties of **1–4**.

2. Materials and Methods

2.1. Chemicals and Reagents

All chemicals in this article were received from Sigma-Aldrich Chemical Company (St. Louis, MI, USA) with pure grade.

2.2. Synthesis of Tosylates **1–4**

General synthetic route for **1–4** was published by Yoshida [18] and used as is. Tosyl chloride (2.1 g, 11 mmol) in dichloromethane (20 mL) was added to a stirred suspension of an fluorinated alcohol $HOCH_2CF_2CF_3$ and $HOCH_2(CF_2)_nCHF_2$ (n = 1, 5, 7) (10 mmol), KOH (0.84 g, 15 mmol), triethylamine (10 mg, 0.1 mmol) and trimethylamine hydrochloride (0.1 g, 1 mmol) in dichloromethane (20 mL) at 5–10 °C, and the mixture was stirred for 1 h and at room temperature for 3–5 h. 20 mL aqueous 1 M hydrochloric acid solution was added to the mixture, the organic layer was separated, washed once with 20 ml of water and dried with anhydrous sodium sulfate. The precipitate was filtered off; the solvent was evaporated on a rotor, obtaining the desired tosylates. The melting points were identical to published data: **1** (52–53 [19]), **2** (13–15 [20]), **3** (34–35 [21]), and **4** (43–44 °C [20]). The yields for **1–4** are equal to 80, 87, 95, and 95%, respectively. The crystals suitable for X-ray analysis were obtained from reaction mass (**1**, **2a**, and **4**) and grown from liquid samples (**2b**, **3a**, and **3b**).

2.3. Single Crystal Structure Analysis

Single crystal X-ray studies of **1–4** were carried out in Center for molecule composition studies of INEOS RAS using Bruker APEX II and Bruker APEX DUO diffractometers. All crystal samples were colorless crystals with low melting point. To prevent damage of the samples and decrease of thermal movement of atoms the measurements were carried out at 120 K.

The structures were solved by direct method and refined in anisotropic approximation for non-hydrogen atoms. Hydrogens atoms of methyl, methylene and aromatic fragments were calculated according to those idealized geometry and refined with constraints applied to C-H bond lengths and equivalent displacement parameters ($U_{iso}(H) = 1.2U_{eq}(C)$ for CH_2, and CH; $U_{iso}(H) = 1.5U_{eq}(C)$ for CH_3 group. All structures were solved with the ShelXT [22] program and refined with the ShelXL [23] program. Molecular graphics was drawn using OLEX2 [24] program. The structure **3a** was refined as inversion twin using TWIN and BASF instructions (Flack parameter is equal to 0.11(17)). CCDC 1907454-1907459 and Table S1 contain the supplementary crystallographic data for **1–4**. These data can be obtained free of charge from The Cambridge Crystallographic Data Centre via https://www.ccdc.cam.ac.uk/structures.

2.4. Quantum Chemical Calculations

All DFT calculations were performed within the PBE exchange-correlation functional using VASP 5.4.1 [25–28]. Atomic coordinates were optimized; however, cell parameters were fixed at their experimental values to prevent cell contraction or expansion (total energies are summarized in Table S2, optimized coordinates can be found in of electronic supplementary information (VASP calculation output section)). To improve the description of van-der-Waals interactions D3 correction [29] was applied. Atomic cores were described using PAW potentials. Valence electrons ($2s$ and $2p$ for O and N atoms; $3p$, and $3s$ for S; $1s$ for H) were described in terms of a plane-wave basis set (the kinetic energy cutoff was at 800 eV). VASP is supplied with library of small-core PAW potentials. Thus, the problems with topological analysis due to usage of pseudopotentials was avoided for intermolecular interactions. The electron density function suitable for analysis in terms of QTAIM theory was obtained in separate single-point calculations of the optimized structures of **1–4** using the fast Fourier transform grid that was twice as dense as the default values (the distances between points in direct space were ~0.03 Å). Topological analysis of electron density in terms QTAIM was carried out with AIM program (a part of ABINIT code [30]). NCI analysis was performed using CRITIC2 software [31].

The energies of intermolecular interactions were evaluated using Espinosa, Mollins and Lecomte correlation formula [32]. The sum of energies of all intermolecular interaction can be associated with the values of lattice energy. In addition to topological analysis, lattice energies were obtained using energy decomposition procedure implemented into CrystalExplorer17 program [33]. The latter approach used experimental X-ray coordinates, while all bonds with hydrogen atoms were normalized to value from neutron diffraction studies. In contrast to VASP calculations, Crystal Explorer used localized basis set 6–31G(d,p) and B3LYP functional. Calculated energies were scaled to account counterpoise and dispersion corrections.

3. Results and Discussion

3.1. Geometry and Crystal Packing of **1–4**

General views of molecules **1–4** are presented at Figures 1–6, while the information about the most important structural parameters is summarized in Table 1 (in addition, molecular structures of **2b** and **3b** are shown at Figures S1 and S2 in supplementary). All compounds crystallized in monoclinic cell. Asymmetric unit of **2b** contains two molecules denoted as **A** and **B**. Other structures are characterized by Z = 1. Bond lengths in tosylate and fluoroalkyl moieties are the same as in the case of similar sulfonates and fluorinated alcohol derivatives in CSD [34]. The length of the terminal C–F and C-C bonds is a bit shorter than in the case of difluoromethylene moieties vicinal to sulfonate ones.

Mutual orientation of a flurorinated alkyl and a tosyl moiety is governed by crystal packing. Torsion angles C1-S1-O1-C8 in **1** and **4** are equal to 71.51(13) and 68.42(13)°, respectively. Conformation of the hydrocarbon chain in polymorph **2a** is almost the same as in molecule **A** of polymorph **2b** (angle C1-S1-O1-C8 is equal to −69.23(12) and −74.894(6)°) while molecule **B** has another conformation (the angles C1-S1-O1-C8 is equal to 77.208(7)°). In polymorphs **3a** and **3b** this angle is equal to −85.8(4) and −69.8(2)°, correspondingly.

Table 1. Selected bond lengths and angles in 1–4.

Structural Parameters (Å and °)	Crystal Structure					
	1	2a	2b	3a	3b	4
S1-O1	1.5980(12)	1.5896(12)	1.5918(18)	1.602(3)	1.586(2)	1.5900(14)
S1-O2	1.4238(14)	1.4304(13)	1.4284(19)	1.438(3)	1.425(2)	1.4347(15)
S1-O3	1.4285(14)	1.4276(13)	1.4273(19)	1.433(4)	1.429(2)	1.4298(15)
S1-C1	1.7474(17)	1.7530(16)	1.749(3)	1.752(5)	1.752(3)	1.7584(19)
O1-C8	1.439(2)	1.445(2)	1.445(3)	1.436(6)	1.438(3)	1.448(2)
C9-F	1.353(2)	1.356(2)	1.358(3)	1.360(5)	1.356(3)	1.357(2)
C-F *	-	-	-	1.348(6)	1.344(3)	1.345(2)
C-F$_{term}$	1.326(2)	1.356(2)	1.354(3)	1.338(8)	1.347(5)	1.350(4)
O1-S1-C1	103.65(8)	103.97(7)	104.04(11)	103.6(2)	103.08(12)	103.32(8)
C8-O1-S1	116.09(11)	116.91(10)	117.77(16)	119.9(3)	116.98(16)	117.07(11)

* - mean C-F distance with exception of vicinal CF$_2$ and terminal CHF$_2$ groups.

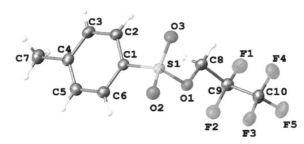

Figure 1. Molecular structure of 1. Atoms are presented as thermal ellipsoids.

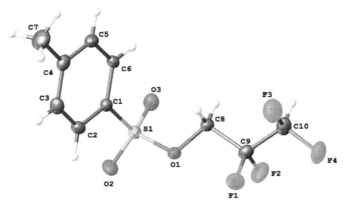

Figure 2. Molecular structure of 2a. Atoms are presented as thermal ellipsoids.

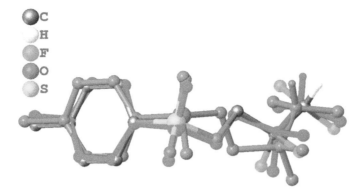

Figure 3. Overlaid molecules in structures **2a** and **2b**. Color code: **2b**, molecule **A**—magenta; **2a**, molecule **B**—blue; and **2a** is colored by element.

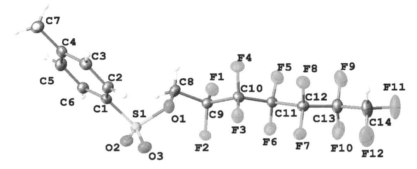

Figure 4. Molecular structure of **3a**. Atoms are presented as thermal ellipsoids.

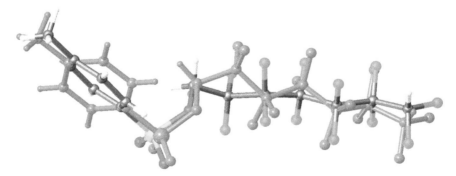

Figure 5. Overlaid molecular structures of polymorphs **3a** (colored by element) and **3b** (blue).

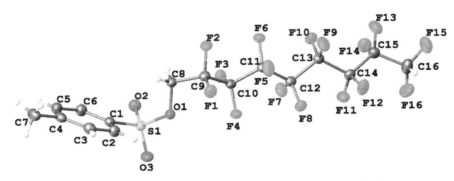

Figure 6. Molecular structure of **4**. Atoms are presented as thermal ellipsoids.

Compounds **1–4** form similar crystal packing, which can be described as a tail-to-tail arrangement of molecules (Figures 7–9). It is noteworthy that the values of *b* side in **1–4** are always equal. Besides, the fluoroalkyl fragments are assembled together, however, there are differences in mutual disposition of molecules in crystal packing. Analysis of short contacts between atoms of these fragments in **1** (Figure 7a) revealed the absence of F ... F contacts (all F ... F distances exceed the sum of those van-der-Waals radii that is equal to 2.9 Å [35]). The most pronounced intermolecular interactions are weak C–H ... O hydrogen bonds and C–H ... π interactions between tosylate moieties. In the case of bulky fluoroalkyl fragments (**2–4**), the F ... F distances became shorter. In several cases these distances are considerably shorter than 2.9 Å (for instance, F3 ... F8[1 − x, −1/2 + Y, 1 − Z] and F7 ... F5[x, −1 + y, z] distances in **3a** and **4** are equal to 2.764(4) and 2.5112(17) Å, respectively). Additional characterization of F ... F contacts using pair C–F ... F–C angles has shown that first angle is close to linear (142–170°), while the second one varies in wide range (106–161°). Generally, the shorter F ... F distance the closer both angles are to 180° that corresponds to halogen-halogen contact of type I according to classification by Desiraju [36]. According to CSD [34], this picture is typical for compounds with polyfluoroalkyl fragments.

Despite the amount of fluorine atoms only few H ... F contacts were found in **1–4**. The strongest contacts are related to the formation of C–H ... F bond with terminal difluoromethyl group. The latter bond can be described as an additional factor that is assisted for arrangement of fluoroalkyl moieties. Thus, the contribution of F ... F to energy of crystalline packing noticeably increases along with the size of fluoroalkane moiety. Unfortunately, the analysis and quantitative estimation of interatomic interactions using only analysis of short contacts is difficult and ambiguous. To provide more information about the role of intermolecular interactions into crystal packing energy quantum chemical calculations were carried out using different DFT functionals and basis sets (PBE/800 eV and CE-B3LYP/6-31G(d,p)).

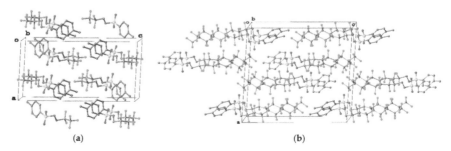

(a) (b)

Figure 7. Crystal packing of **1** (a) and **4** (b).

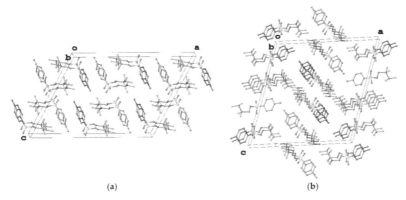

Figure 8. Crystal packing of polymorphs **2a** (a) and **2b** (b).

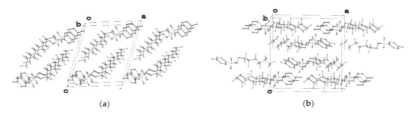

Figure 9. Crystal packing of polymorphs **3a** (a) and **3b** (b).

3.2. Quantum Chemical Calculations of Crystal Structures **1**–**4**

The bond lengths obtained in PBE calculations of crystal structures **1**–**4** and isolated molecules of $(CHF_2)(CF_2)_nCH_2OTs$ is in satisfactory agreement with experimental values. Root-mean-square deviations between experimental and calculated coordinates of non-hydrogen atoms are 0.032 (**1**), 0.035 (**2a**), 0.032 (**2b**, molecule A), 0.038 (**2b**, molecule B), 0.048 (**3a**), 0.055 (**3b**), and 0.043(**4**) Å. The differences in values of S-O, S-C and C-F bonds are 0.01–0.03 Å. The C-H bonds are elongated up to 0.12 Å, however, it is expected because the coordinates of hydrogen atoms cannot be measured with sufficient accuracy using X-ray diffraction. Intermolecular distances between non-hydrogen atoms somewhat changed, as compared to X-ray structures of studied compounds. Fortunately, the deviation between the calculated and the experimental structure are not so pronounced, therefore we can expect that the values related to the energies of intermolecular interactions from PBE-D3 calculations are valid for analysis of crystal packing.

Among the computational methods for analysis of intermolecular interactions the most popular and informative approach is R. Bader's quantum theory of "Atoms in Molecules" (QTAIM) [37]. According to QTAIM any intermolecular contact can be detected by topological analysis of electron density function calculated for non-periodic (molecules or molecular associates) and periodic systems like crystals or surfaces. Analysis of calculated electron density in terms of QTAIM has shown that bond critical points (*bcp*) were found for all expected covalent bonds. The *bcp*s related to bonds formed by sulfur atoms is characterized by positive value of Laplacian of $\rho(\mathbf{r})$ ($\nabla\rho(\mathbf{r})$) and negative one of local energy density ($H^e(\mathbf{r})$) that indicate its highly polar character. The rest of bonds in studied structures can be described as typical covalent ones because the values of $\nabla\rho(\mathbf{r})$ and $H^e(\mathbf{r})$ are negative.

Intermolecular interactions found by QTAIM analysis correspond to closed shell interactions (positive values of $\nabla\rho(\mathbf{r})$ and $H^e(\mathbf{r})$ in corresponding *bcp*s). It is noteworthy, that some types of interactions are not possible to detect on the base of structure analysis. For instance, QTAIM analysis revealed the presence of H … H, C–H … π and F … π interactions between methylene, phenyl and fluoroalkyl groups. The strongest intermolecular bonds are C–H ... O hydrogen bonds between difluoromethyl and sulfonate moieties. Their energies raised along with the size of fluoroalkyl fragment from 5.8 kJ/mol in the case of CF_2CHF_2 group (**2a**) up to 12.3 kJ/mol in **4** ((CF_2)$_7CHF_2$). In **2b** the hydrogen atom of CHF_2 group in molecule **A** formed bifurcate C–H ... O and C–H ... F bonds (8.7 and 5.9 kJ/mol) with sulfonate group of adjacent molecule **A** and CHF_2 one of molecule **B**. The hydrogen atom of CHF_2 group in molecule **B** participates only in C–H ... F bond (4.9 kJ/mol) with difluoromethyl moiety of molecule **A**. In polymorphs of $TsOCH_2(CF_2)_5CHF_2$ the energies of present C–H ... F bond differ by several kJ/mol (10.1 and 13.2 in **3a** and **3b**, respectively). In **1** where CHF_2 group is changed to CF_3 one the strongest intermolecular interaction is the C–H ... O bond between a phenyl ring and a sulfonate moiety (7.8 kJ/mol). As rule, C–H ... F bonds are somewhat weaker than C–H ... O ones, the strongest hydrogen bonds of such type do not exceed 8 kJ/mol. Few *bcps* were also found for F ... C contacts that mainly correspond to F ... π interaction between a fluorine atom and a phenyl group. In addition, F ... O interactions were detected. Two latter types of interactions are very weak (less than 2 kJ/mol).

Bcps related to F ... F contacts attract especial interest. Some of fluorine atoms are not involved in F ... F interactions, while the others form up to four intra- and intermolecular interactions of this type. Intramolecular F ... F interactions were revealed for structure **3a** (Table 2). In **3b** that is another conformer (Figure 5) and polymorph of $TsOCH_2(CF_2)_5CHF_2$ such strong intramolecular interactions between fluorine atoms were not found. In fact, numerous F ... F interactions formed a framework responsible for arrangement of fluoroalkyl fragments. In contrast to F ... C and F ... O interactions the energies of F ... F vary in wide range from 0.5 to 13 kJ/mol. The strongest F ... F interactions (for instance F5 ... F7 in **4**) appeared to be stronger than C–H ... O hydrogen bonds. According to literature, the analysis of valence electron density or deformation electron density ($\Delta\rho$) distribution in the region of shortest F ... F contacts clearly demonstrated that a lone pair of fluorine atoms is directed toward the local depletion of electron density between electron pairs of another fluorine atom [38] ("peak-hole interaction"). QTAIM and $\Delta\rho$ maps are very comprehensive tools for unexperienced reader, however, these methods cannot be used to analyze the entire region related to F ... F interaction. Complementary information on intermolecular bonding was obtained using NCI (non-covalent interaction) method [39,40] based on dimensionless RDG (reduced density gradient) function related to the magnitude of λ_2 eigenvalue(sign$\lambda_2$rho). To make analysis interatomic interactions more comprehensive 3D isosurfaces of RDG function in the regions of these interactions were colored according to the sign of λ_2 multiplied by $\rho(r)$. Similarly to $\Delta\rho$, NCI method can be used as indicator for redistribution of electron density as result of chemical bond formation. Moreover, NCI method is much more useful for weak intermolecular interactions than $\Delta\rho$. Maxima of RDG can be described as analog of *bcp*, those presence in interatomic region is an indicative for corresponding interaction. The shape and the volume of above-mentioned maxima supply additional information on interatomic interactions. Additionally, it is important to analyze the sign of λ_2. As rule, the maxima for rather strong intermolecular interactions like classic hydrogen bonds (O–H ... O or N–H ... O) are small and they have discoidal shape. The sign of λ_2 is mainly negative that is an indicative for attractive nature of classic hydrogen bonds. On the contrary, the maxima for weak H ... H interactions are characterized by rather large area and they had no definite shape. At the same time, the regions with positive sigh of λ_2 are dominated over those with a negative sign of λ_2.

Table 2. Strongest F . . . F interactions in **1–4** estimated using the EML [32] correlation.

Interactions (Compound)	Type (*Intramolecular*/Intermolecular)	Experimental Distance	Calculated Distance	Energy, kJ/mol
F4 . . . F4 (**1**)	intermolecular	2.978(3)	2.925	−4.2
F3 . . . F3 (**2a**)	intermolecular	2.750(2)	2.785	−5.9
F4 . . . F4A (**2b**)	intermolecular	3.074(2)	3.046	−3.1
F3 . . . F7 (**3a**)	*intramolecular*	2.633(5)	2.649	−12.2
F3 . . . F8 (**3a**)	intermolecular	2.764(4)	2.785	−5.5
F5 . . . F9 (**3a**)	*intramolecular*	2.582(5)	2.620	−4.6
F4 . . . F7 (**3b**)	intermolecular	2.921(2)	2.876	−4.6
F5 . . . F8 (**3b**)	intermolecular	2.543(3)	2.572	−10.5
F4 . . . F11 (**4**)	intermolecular	2.9031(17)	2.937	−3.8
F4 . . . F12 (**4**)	intermolecular	2.7953917)	2.777	−5.5
F5 . . . F7 (**4**)	intermolecular	2.5112(17)	2.507	−13.0
F14 . . . F14 (**4**)	intermolecular	2.942(3)	2.895	−5.0

Since a lot of F . . . F interactions were found in **1–4** it is important to analyze these interactions using NCI to find similarities and differences between weakest and strongest ones related to their nature. It is clear (Figures 9–11) that strongest intra- and intermolecular F . . . F interactions are characterized by negative values of λ_2 similarly to hydrogen bond C-H . . . O between sulfonate group and terminal CHF_2 group. These regions are highlighted by blue or light blue color on Figures 10–12. Thus, F . . . F interactions shown in Table 2 can be described as mostly attractive ones. At the same time, the values of sign$\lambda_2$rho for the majority of intermolecular F . . . F interactions are close to zero and the sign of λ_2 varied from positive to negative (green color on Figures 10–12), so it is very difficult to unambiguously describe them as attractive or repulsive.

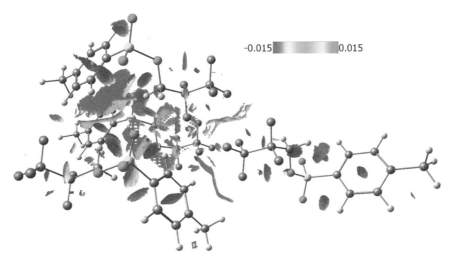

−0.015 ▬▬▬▬▬ 0.015

Figure 10. 3D surface of RDG (0.6 a.u.) colored according to sign(λ_2)ρ function in **1** illustrating the interaction between CF_3 groups.

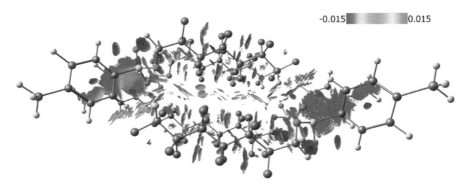

-0.015 ▮▮▮▮▮▮ 0.015

Figure 11. 3D surface of RDG (0.6 a.u.) colored according to sign (λ_2) ρ function in **3a** illustrating the interaction between $(CF_2)_5CHF_2$ groups. Intermolecular interactions between fluorine atoms are shown at middle bottom.

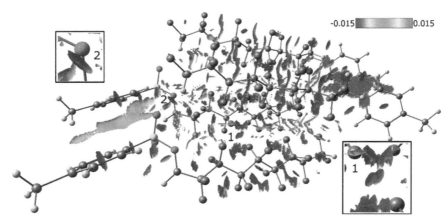

-0.015 ▮▮▮▮▮▮ 0.015

Figure 12. 3D surface of RDG (0.6 a.u.) colored according to sign (λ_2) ρ function in **4** illustrating the interactions between $(CF_2)_5CHF_2$ groups. Intermolecular C-H . . . O bond (1) and strong F . . . F interaction (2) are shown.

3.3. Lattice Energies and the Role of F . . . F Interactions

The energies of intermolecular interactions calculated from ρ(**r**) provided the opportunity to qualitatively estimate the contribution of F . . . F ones to the energy of crystal packing. In fact, the latter value is the sum of the energies of all intermolecular interactions found. This way to calculate the energy related to molecular association is very attractive, however, there is at least one serious drawback. This problem is related to empirical character of EML correlation formula, so it was criticized by Spackman [41]. According to Reference [41] EML formula in most cases underestimated the energy of intermolecular interactions by substantial amount as compared to the method implemented to CrystalExplorer program (CE-B3LYP). Severe judgement about EML correlation were expressed in the paper by Kuznetsov [42]. On the other hand, according to other published articles the lattice energies calculated from EML correlation provided reasonable values that agreed with experimental sublimation heat [43–45]. Thus, the reference method for estimation of lattice energies is necessary to attest the results of QTAIM approach and EML correlation formula for compounds with fluorinated alkyl moieties. The CE-B3LYP method seems to be the most reliable and comprehensive method for calculation of intermolecular potentials available for crystallographers. As result of CE-B3LYP calculations the values of interactions of a target molecule with its neighbors in molecular cluster generated according to space group symmetry operations are provided. The calculation of the intermolecular energies demonstrated the similarity of crystal packing motifs in **1–4** (See supporting information (CrystalExplorer17 output) for details). The strongest intermolecular interactions are observed between fluoroalkyl fragments. It is clear that energies of above interactions are increased along with the size of fluoroalkyl fragments. The value of lattice energy can be easily calculated from the data on all intermolecular interactions in cluster. Unfortunately, we encountered unexpected work of CrystalExplorer17 in the case of two independent molecules (**2b**). It was impossible to calculate lattice energy for two independent molecules separately. The value obtained for two independent molecules as a whole (−173.8 kJ/mol) is not reliable because the interactions between them are neglected.

The information about lattice energies estimated from QTAIM/EML method and CE-B3LYP/6-31G(d,p) calculations is presented in Figures 12 and 13.

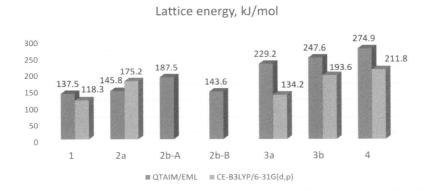

Figure 13. Calculated lattice energies in **1–4**. The values are shown above the bars. Molecules A and B denoted as **2b**-A and **2b**-B. The values were multiplied by −1.

It can be seen from Figure 13 that QTAIM/EML overestimated the lattice energy in all structures except for **2a**. Nevertheless, both methods predicted that polymorph **3b** is more stable than **3a**. This result was also verified by comparison of total energies of **3a** and **3b** (total energy of the latter divided by Z is 2.47 kJ/mol larger than the former). The situation with polymorphs **2a** and **2b** is the same as in **3a** and **3b**. Molecules **A** and **B** have noticeably different values of lattice energy. If averaged value of lattice energies of molecules A and B (165.6 kJ/mol) was taken as measure of stability, then **2b** is appeared to be more stable than **2a**. The comparison of the total energies of **2a** and **2b** divided by Z also supports this conclusion (the difference is 1.35 kJ/mol). Total contributions related to the most prominent intermolecular interactions are shown at Figure 14. It is logically to assume that contribution of F ... F interactions to lattice energy will increase along the amount of fluorine atoms, while the contribution of H ... O interactions (namely C-H ... O hydrogen bonds) will decrease. At first glance, the results of QTAIM/EML evaluations agree with this assumption but there are two exceptions. The first one is related to polymorphism, because the contribution of F ... F interactions can vary due to way of molecular packing. The part related to F ... F in **3a** exceeds that in **3b**, although, their absolute values are almost equal (78.4 and 76.2 kJ/mol). At the same time, the percentage of H ... O interactions in **3a** is less than in **3b**. It is necessary to remind that polymorph **3b** is more stable than **3a**. Again, the situation with polymorphs **2a** and **2b** is the same. The contribution of F ... F interactions to lattice energy in **2b** is larger than in **2a**. At the same time, the percentage of H ... O and H ... F interactions in **2b** is considerably larger than in **2a** case. Thus, H ... O interactions can be the main factor that made polymorphs **2b** and **3b** more favorable than **3a** and **3a** ones. This conclusion is in agreement with recent paper by Saha [46]. The second exception is related to contribution of O . . . H interactions in the case of **1**. There is no terminal CHF_2 group in molecule of $PhO_2SOC_2F_5$, so this group cannot participate in C–H . . . O bonds with phenyl group that explain so low contribution of H . . . O interactions. Various interactions formed by hydrogen atoms (C–H . . . π, H . . . H, H . . . C, denoted as "other" on Figure 14) are responsible for more than 30% of lattice energy in **1**.

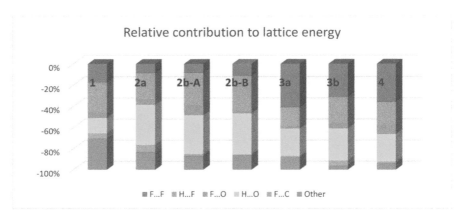

Figure 14. Relative contribution of various intermolecular interactions to lattice energy according QTAIM/EML.

4. Conclusions

The analysis of intermolecular interactions has shown that the contribution of interactions formed by fluorine atoms almost linearly depends on its amount. All studied compounds contain only three oxygen atoms, however, the role of C-H . . . O bonds is prominent even in the case of **4** that contain sixteen fluorine atoms. The contribution of F . . . F interaction does not exceed 40% even in the case of **4**. Possibly such a relatively small contribution of F . . . F interaction is related to its specific nature. Indeed, according to the results NCI and QTAIM analysis F . . . F interactions in **1–4** hardly can be described as attractive as weak hydrogen bonds. Indeed, there are several strong interactions of such type with apparently attractive character, however, their total energy is rather small as compared to those for analogous weak interactions. The comparison of lattice energies calculated for polymorphs **2a**, **3a**, **2b** and **3b** has shown that increase of F . . . F contribution do not lead to stabilization of crystal packing in contrast to intermolecular C-H . . . O that have mostly attractive nature. In other words, the lower contribution of F . . . F interactions to the total energy, the more stable a polymorph is.

Supplementary Materials: The following are available online at http://www.mdpi.com/2073-4352/9/5/242/s1, Crystallographic data: Table S1 Crystallographic data for **1–4**, Figure S1. Molecular structure of **2b**, Figure S2. Molecular structure of **3b**. VASP output: Parameters of unit cell and optimized fractional coordinates, Table S2: Total energy of unit cell (OUTCAR file). Output of Crystal Explorer program, References.

Author Contributions: Conceptualization, writing—original draft preparation, review and editing—D.E.A. and A.A.K.; synthesis—A.V.L. and D.E.A.; X-ray diffraction studies and quantum chemical calculations—A.D.V.

Funding: Quantum chemical studies and the synthesis of **2–4** were supported by Russian Science Foundation (grant 18-73-00339). The synthesis of compound **1** was supported by Russian Foundation for Basic Research (grant no. 18-29-20102).

Acknowledgments: X-ray diffraction studies were performed with the support from Ministry of Science and Higher Education of the Russian Federation using the equipment of Center for molecule composition studies of INEOS RAS.

References

1. Ahrens, L.; Bundschuh, M. Fate and effects of poly- and perfluoroalkyl substances in the aquatic environment: A Review: Fate and effects of polyfluoroalkyl and perfluoroalkyl substances. *Environ. Toxicol. Chem.* **2014**, *33*, 1921–1929. [CrossRef] [PubMed]

2. Ospinal-Jiménez, M.; Pozzo, D.C. Structural analysis of protein denaturation with alkyl perfluorinated sulfonates. *Langmuir* **2012**, *28*, 17749–17760. [CrossRef]

3. Rosholm, K.R.; Arouri, A.; Hansen, P.L.; González-Pérez, A.; Mouritsen, O.G. Characterization of fluorinated catansomes: A promising vector in drug-delivery. *Langmuir* **2012**, *28*, 2773–2781. [CrossRef] [PubMed]

4. Krafft, M. Fluorocarbons and fluorinated amphiphiles in drug delivery and biomedical research. *Adv. Drug Deliv. Rev.* **2001**, *47*, 209–228. [CrossRef]

5. Massi, L.; Guittard, F.; Géribaldi, S.; Levy, R.; Duccini, Y. Antimicrobial properties of highly fluorinated bis-ammonium salts. *Int. J. Antimicrob. Agents* **2003**, *21*, 20–26. [CrossRef]

6. Luo, Z.; Shi, X.; Hu, Q.; Zhao, B.; Huang, M. Structural evidence of perfluorooctane sulfonate transport by human serum albumin. *Chem. Res. Toxicol.* **2012**, *25*, 990–992. [PubMed]

7. Du, Z.; Deng, S.; Bei, Y.; Huang, Q.; Wang, B.; Huang, J.; Yu, G. Adsorption behavior and mechanism of perfluorinated compounds on various adsorbents—A Review. *J. Hazard. Mater.* **2014**, *274*, 443–454. [CrossRef] [PubMed]

8. Wadekar, M.N.; Abezgaus, L.; Djanashvili, K.; Jager, W.F.; Mendes, E.; Picken, S.J.; Dganit, D. Supramolecular "leeks" of a fluorinated hybrid amphiphile that self-assembles into a disordered columnar phase. *J. Phys. Chem. B* **2013**, *117*, 2820–2826. [CrossRef] [PubMed]

9. Prasanna, M.D.; Row, T.N.G. Weak interactions involving organic fluorine: analysis of structural motifs in flunazirine and haloperidol. *J. Mol. Struct.* **2001**, *562*, 55–61. [CrossRef]

10. Choudhury, A.R.; Urs, U.K.; Guru Row, T.N.; Nagarajan, K. Weak interactions involving organic fluorine: a comparative study of the crystal packing in substituted isoquinolines. *J. Mol. Struct.* **2002**, *605*, 71–77. [CrossRef]

11. Osuna, R.M.; Hernández, V.; Navarrete, J.T.L.; D'Oria, E.; Novoa, J.J. Theoretical evaluation of the nature and strength of the F···F intermolecular interactions present in fluorinated hydrocarbons. *Theor. Chem. Acc.* **2011**, *128*, 541–553. [CrossRef]

12. Nayak, S.K.; Reddy, M.K.; Row, T.N.G.; Chopra, D. Role of hetero-halogen (F···X, X = Cl, Br, and I) or homo-halogen (X···X, X = F, Cl, Br, and I) interactions in substituted benzanilides. *Cryst. Growth Des.* **2011**, *11*, 1578–1596. [CrossRef]

13. Prakash, G.K.S.; Wang, F.; Rahm, M.; Shen, J.; Ni, C.; Haiges, R.; Olah, G.A. On the nature of CH···FC interactions in hindered CF3C(sp3) bond rotations. *Angew. Chem. Int. Ed.* **2011**, *50*, 11761–11764. [CrossRef]

14. Alkorta, I.; Elguero, J. Fluorine–fluorine interactions: NMR and AIM analysis. *Struct. Chem.* **2004**, *15*, 117–120. [CrossRef]

15. Levina, E.O.; Chernyshov, I.Y.; Voronin, A.P.; Alekseiko, L.N.; Stash, A.I.; Vener, M.V. Solving the enigma of weak fluorine contacts in the solid state: A periodic dft study of fluorinated organic crystals. *RSC Adv.* **2019**, *9*, 12520–12537. [CrossRef]

16. Omorodion, H.; Twamley, B.; Platts, J.A.; Baker, R.J. Further evidence on the importance of fluorous–fluorous interactions in supramolecular chemistry: a combined structural and computational study. *Cryst. Growth Des.* **2015**, *15*, 2835–2841. [CrossRef]

17. Baker, R.J.; Colavita, P.E.; Murphy, D.M.; Platts, J.A.; Wallis, J.D. Fluorine–fluorine interactions in the solid state: an experimental and theoretical study. *J. Phys. Chem. A* **2012**, *116*, 1435–1444. [CrossRef] [PubMed]

18. Yoshida, Y.; Sakakura, Y.; Aso, N.; Okada, S.; Tanabe, Y. Practical and efficient methods for sulfonylation of alcohols using Ts(Ms)Cl/Et$_3$N and Catalytic Me$_3$H·HCl as combined base: Promising alternative to traditional pyridine. *Tetrahedron* **1999**, *55*, 2183–2192. [CrossRef]

19. Banerjee, S.; Vidya, V.M.; Savyasachi, A.J.; Maitra, U. Perfluoroalkyl bile esters: A new class of efficient gelators of organic and aqueous–Organic media. *J. Mater. Chem.* **2011**, *21*, 14693–14705. [CrossRef]

20. Cohen, W.V. Nucleophilic substitution in fluoroalkyl sulfates, sulfonates, and related compounds. *J. Org. Chem.* **1961**, *26*, 4021–4026. [CrossRef]

21. Shilin, S.; Florensova, O.; Chernov, N.; Voronkov, M. Alpha, alpha, omega-trihydro-alpha-halo perfluoroalkanes. *Zhurnal Obshchei Khimii* **1991**, *61*, 1838–1840.

22. Sheldrick, G.M. SHELXT—Integrated space-group and crystal-structure determination. *Acta Crystallogr.* **2015**, *A71*, 3–8. [CrossRef]
23. Sheldrick, G.M. Crystal structure refinement with SHELXL. *Acta Crystallogr.* **2015**, *C71*, 3–8.
24. Dolomanov, O.V.; Bourhis, L.J.; Gildea, R.J.; Howard, J.A.K.; Puschmann, H. OLEX2: A complete structure solution, refinement and analysis program. *J. Appl. Crystallogr.* **2009**, *42*, 339–341. [CrossRef]
25. Kresse, G.; Hafner, J. Ab initio molecular dynamics for liquid metals. *Phys. Rev. B* **1993**, *47*, 558. [CrossRef]
26. Kresse, G.; Hafner, J. Ab initio molecular-dynamics simulation of the liquid-metal amorphous-semiconductor transition in germanium. *Phys. Rev. B* **1994**, *49*, 14251–14269. [CrossRef]
27. Kresse, G.; Furthmuller, J. Efficient iterative schemes for ab initio total-energy calculations using a plane-wave basis set. *Phys. Rev. B* **1996**, *54*, 11169. [CrossRef]
28. Kresse, G.; Furthmuller, J. Efficiency of ab-initio total energy calculations for metals and semiconductors using a plane-wave basis set. *Comput. Mater. Sci.* **1996**, *6*, 15–50. [CrossRef]
29. Moellmann, J.; Grimme, S. Importance of London dispersion effects for the packing of molecular crystals: A case study for intramolecular stacking in a Bis-thiophene derivative. *Phys. Chem. Chem. Phys.* **2010**, *12*, 8500. [CrossRef] [PubMed]
30. Gonze, X.; Beuken, J.-M.; Caracas, R.; Detraux, F.; Fuchs, M.; Rignanese, G.-M.; Sindic, L.; Verstraete, M.; Zerah, G.; Jollet, F.; et al. First-principles computation of material properties: The ABINIT software project. *Comput. Mater. Sci.* **2002**, *25*, 478–492. [CrossRef]
31. Otero-de-la-Roza, A.; Johnson, E.R.; Luaña, V. Critic2: A program for real-space analysis of quantum chemical interactions in solids. *Comput. Phys. Commun.* **2014**, *185*, 1007–1018. [CrossRef]
32. Espinosa, E.; Molins, E.; Lecomte, C. Hydrogen bond strengths revealed by topological analyses of experimentally observed electron densities. *Chem. Phys. Lett.* **1998**, *285*, 170–173. [CrossRef]
33. Mackenzie, C.F.; Spackman, P.R.; Jayatilaka, D.; Spackman, M.A. CrystalExplorer model energies and energy frameworks: Extension to metal coordination compounds, organic salts, solvates and open-shell systems. *IUCrJ* **2017**, *4*, 575–587. [CrossRef]
34. Groom, C.R.; Bruno, I.J.; Lightfoot, M.P.; Ward, S.C. The Cambridge structural database. *Acta Crystallogr. Sect. B Struct. Sci. Cryst. Eng. Mater.* **2016**, *72*, 171–179. [CrossRef]
35. Rowland, R.S.; Taylor, R. Intermolecular nonbonded contact distances in organic crystal structures: comparison with distances expected from van der Waals Radii. *J. Phys. Chem.* **1996**, *100*, 7384–7391. [CrossRef]
36. Desiraju, G.R.; Parthasarathy, R. The nature of Halogen...Halogen interactions: Are short halogen contacts due to specific attractive forces or due to close packing of nonspherical atoms? *J. Am. Chem. Soc.* **1989**, *111*, 8725–8726. [CrossRef]
37. Bader, R.W.F. *Atoms in Molecules: A Quantum Theory*; Oxford University Press: New York, NY, USA, 1990.
38. Levin, V.V.; Dilman, A.D.; Korlyukov, A.A.; Belyakov, P.A.; Struchkova, M.I.; Antipin, M.Yu.; Tartakovsky, V.A. Synthesis and structures of tris(pentafluorophenyl)silylamines. *Russ. Chem. Bull.* **2007**, *56*, 1394–1401. [CrossRef]
39. Johnson, E.R.; Keinan, S.; Mori-Sánchez, P.; Contreras-García, J.; Cohen, A.J.; Yang, W. Revealing noncovalent interactions. *J. Am. Chem. Soc.* **2010**, *132*, 6498–6506. [CrossRef]
40. Saleh, G.; Gatti, C.; Lo Presti, L. Non-covalent Interaction via the reduced density gradient: independent atom model vs experimental multipolar electron densities. *Comput. Theor. Chem.* **2012**, *998*, 148–163. [CrossRef]
41. Spackman, M.A. How reliable are intermolecular interaction energies estimated from topological analysis of experimental electron densities? *Cryst. Growth Des.* **2015**, *15*, 5624–5628. [CrossRef]
42. Kuznetsov, M.L. Can halogen bond energy be reliably estimated from electron density properties at bond critical point? The case of the $(A)_nZ—Y\bullet\bullet\bullet X - (X, Y = F, Cl, Br)$ interactions. *Int. J. Quantum Chem.* **2019**, *119*, e25869. [CrossRef]
43. Nelyubina, Y.V.; Korlyukov, A.A.; Lyssenko, K.A. Probing weak intermolecular interactions by using the invariom approach: A comparative study of s-tetrazine. *Chem. Eur. J.* **2014**, 6978–6984. [CrossRef] [PubMed]
44. Nelyubina, Y.V.; Korlyukov, A.A.; Lyssenko, K.A. Probing systematic errors in experimental charge density by multipole and invariom modeling: A twinned crystal of 1,10-phenanthroline hydrate. *Mendeleev Commun.* **2014**, *24*, 286–289. [CrossRef]

45. Lyssenko, K.A.; Korlyukov, A.A.; Golovanov, D.G.; Ketkov, S.Yu.; Antipin, M.Yu. Estimation of the barrier to rotation of benzene in the $(\eta^6\text{-}C_6H_6)_2Cr$ crystal via topological analysis of the electron density distribution function. *J. Phys. Chem. A* **2006**, *110*, 6545–6551. [CrossRef]

46. Saha, A.; Rather, S.A.; Sharada, D.; Saha, B.K. 'C–X···X–C vs C–H···X–C, which one is the more dominant interaction in crystal packing (X = Halogen)? *Cryst. Growth Des.* **2018**, *18*, 6084–6090. [CrossRef]

Review

Organoelement Compounds Crystallized In Situ: Weak Intermolecular Interactions and Lattice Energies

Alexander D. Volodin *, Alexander A. Korlyukov and Alexander F. Smol'yakov

A. N. Nesmeyanov Institute of Organoelement Compounds, Russian Academy of Science, 28 Vavilova str., Moscow 119991, Russia; alex@xrlab.ineos.ac.ru (A.A.K.); smolyakov@ineos.ac.ru (A.F.S.)
* Correspondence: alex.d.volodin@gmail.com or volodin@ineos.ac.ru

Received: 20 November 2019; Accepted: 26 December 2019; Published: 31 December 2019

Abstract: The in situ crystallization is the most suitable way to obtain a crystal of a low-melting-point compound to determine its structure via X-Ray diffraction. Herein, the intermolecular interactions and some crystal properties of low-melting-point organoelement compounds (lattice energies, melting points, etc.) are discussed. The discussed structures were divided into two groups: organoelement compounds of groups 13–16 and organofluorine compounds with other halogen atoms (Cl, Br, I). The most of intermolecular interactions in the first group are represented by weak hydrogen bonds and H$\cdots$H interactions. The crystal packing of the second group of compounds is stabilized by various interactions between halogen atoms in conjunction with hydrogen bonding and stacking interactions. The data on intermolecular interactions from the analysis of crystal packing allowed us to obtain correlations between lattice energies and Hirshfeld molecular surface areas, molecular volumes, and melting points.

Keywords: X-ray crystallography; in situ crystallization; Hirshfeld surface analyzes; lattice energies; intermolecular interactions; packing motifs; polymorph stability

---

1. Introduction

X-ray diffraction remains the most comprehensive and reliable method available for the studies of the structure and geometry of small molecules and their crystal structures, though numerous successful investigations have been performed with other methods, such as neutron and electron diffraction. Organic compounds that are liquids at room temperature and standard pressure are a special case.

In general, the factors affecting a decrease in melting point seem obvious: small molecular volume, high flexibility of molecular fragments (low barrier of rotation around chemical bonds, low deformation barrier of bond angles, etc.), and weak forces responsible for intermolecular bonding between molecules. As a consequence, such compounds crystallize at temperatures lower than rigid molecules with strong intermolecular interactions. In many cases, a disorder of molecular fragments (or even whole molecules) is observed. The structural data obtained from the corresponding diffraction measurements suffer from many uncertainties related to poor completeness and/or low resolution of the datasets. Nevertheless, it is hard to overestimate the importance of information obtained for the species with low melting points. Indeed, in the case of many low-melting-point compounds, X-ray diffraction is the only way to determine crystal and molecular structure, evaluate the effects of crystal packing, and evaluate the strength of intermolecular interactions. Unfortunately, the methods of electron and neutron diffraction cannot be applied due to a difficult experimental setup or thermal degradation. This statement is more valid for gases and their cocrystals (including gas hydrates [1,2] and gas-liquid systems like acetone-acetylene [3]) rather than for ordinary liquid substances. The number of papers dedicated to the structure determination of samples that are liquid at room temperature has increased gradually due to the wide implementation of methods related to in situ crystallization.

The structural data for compounds that are liquids at room temperature are summarized in a number of reviews [4–7]. In these publications, the most attention was paid to organic [4,5] and bioorganic [6,7] compounds, while the reviews focused on the crystal structures of organoelement compounds that are liquids at room temperature are very rare. Indeed, organoelement compounds are usually more challenging samples for crystallization due to difficulties with purification, air and moisture sensitivity, and relatively high viscosity. Sometimes, the similarities between the crystal structures of organoelement compounds and the structures of organic ones can be observed [8]. Bearing in a mind the practical significance of many organoelement compounds (owing to their importance by themselves or as precursors for the production of catalysts, drugs, and polymers), we can affirm that a review dedicated to the analysis of the crystal structures of these substances would be useful. This work is focused on the analysis of the structures, crystal packing, and intermolecular interactions in organic compounds containing Si, Ge, P, S, and halogen atoms. To analyze the strength of intermolecular interactions in the crystals of compounds, we chose the method of energy frameworks introduced by Spackman [9] because this information is usually unavailable in original papers. The latter method is based on quantum chemical calculations of pairwise interactions in molecular clusters generated from atomic coordinates according to symmetry operations utilizing special corrections to reduce computational errors. These calculations can be done on a standard PC and do not require a considerable computational effort. The results of such a computational evaluation of intermolecular interactions were compared with physical parameters (melting points) or molecular characteristics (molecular volumes and areas).

Our paper is divided into sections dedicated to elements of various groups. It begins with a short historical overview dedicated to the development of in situ crystallization techniques. The final part contains the corresponding analysis of the calculated lattice energies and the nature of intermolecular interactions.

2. In Situ Crystallization: A Retrospective View

The first examples of in situ crystallization were published in the thirties of the past century. The first compounds that underwent in situ crystallization were benzene [10] and cyclohexane [11]. Their melting points are slightly below room temperature, which was crucial at that time, as programmable cooling devices were not available.

Cox E.G. grew a single crystal of benzene in a sealed capsule by cooling with frozen carbon dioxide [12]. The equipment used in this work could not maintain temperatures lower than −40 °C; however, it was sufficient for benzene that started to crystallize at −22 °C in the orthorhombic cell (space group *P*bca). The attempt to study the crystal structure of cyclohexane initially failed. In 1930, Hassel and Kringstad [13] used a Debye camera (FeKα-radiation) and solid CO_2-acetone mixture to crystallize cyclohexane but they could only reveal the cubic symmetry of the unit cell (a ≈ 8.41 Å). Later, Lonsdale and Smith successfully obtained a single crystal of cyclohexane in a preliminarily dried cellophane tube [14] using advanced equipment that allowed them to reach −108 °C. Cellophane tubes are transparent to copper X-ray radiation and are not as fragile as glass capillaries. It was shown that cyclohexane crystallizes in a cubic cell (space group *Fm*3m) and has a chair conformation.

In situ crystallization was applied to verify the idea of the 3c–2e bond in diborane developed by Price [15,16]. At that time, a number of compounds were known that could not be described by Lewis-based valence schemes, whereas MO-LCAO methods were not widely applicable. The explanation of the diborane structure was the cornerstone for the progress in the chemical bond theory, so it was necessary to solve the problem of the diborane structure as soon as possible. Streib and Libscomb developed an original cooling device that used liquid nitrogen and helium to achieve temperatures not far from absolute zero (2–5 K) [17]. The cell used for in situ crystallization had beryllium windows that are transparent to X-rays. The crystals could be obtained from the gas phase upon condensation on the metal rod that, in turn, was connected to a Dewar's vessel. The described device could control the temperature of the rod via a heating coil; however, it was not easy to control the process of crystallization. Fortunately, in 1965, W. Smith and W. Lipscomb successfully grew crystals of the two most thermodynamically stable α and β-phases of diborane [18]. The B–B length in the crystals of these phases was in perfect agreement with the idea of a 3c–2e bond predicted in one of the early MO-LCAO calculations. The same apparatus was used to grow α- and β-phases of nitrogen, β-phase of fluorine, and γ-phase of oxygen [19,20]. These results had great importance because they allowed deriving the first van der Waals radii for fluorine and oxygen, which are still in use for analysis of crystal packing. In 1959, J. Trotter was the first who used a thin-walled capillary to grow single crystals of nitrobenzene [21] using a low-temperature device initially developed by Burbank and Bensey in 1953 [22]. The capillary was filled by liquid nitrobenzene and cooled down to −30 °C, and spontaneous crystallization was observed. This method is not applicable for the majority of liquid samples; however, thin walled glass is still widely used in common practice to grow crystals because it is simple to use and because the length and diameter of the capillary can be varied. Almost perfect single crystals can be grown by combining local cooling and heating, which was first introduced by W. Smith and W. Lipscomb. Unfortunately, it is very difficult to use in the case of samples that are gases at room temperature.

The progress in the in situ crystallization methods in the period from the mid-70 s to mid-90 s was related to the development of new crystallization techniques and hardware. At that point in time, flash freezing with subsequent recrystallization from a melting liquid was adapted for widespread use. Step-by-step local heating of a selected zone used for recrystallization provided an order in crystallites, and as a result, almost perfect single crystals were achieved. In order to control the crystal growth more precisely, the temperature of the heating coil could be controlled by changing the electric current intensity. Movement of the heating coil along the capillary and in the perpendicular direction allowed one to control the temperature gradient and the size of the local heating zone. In the same period of time, power light sources became used as heater elements for local heating instead of heating coils. In 1979, Zimmermann et al. published the scheme of equipment for in situ crystallization in a glass capillary where local heating was achieved using a power LED and elliptic mirrors to focus light on the selected zone. This equipment is suitable for achieving very high temperatures (up to 3000 K) and it can be used even in the absence of gravity, in a space laboratory [23]. Similar equipment for in situ crystallization inside the cabinet of an X-ray diffractometer was applied by Boese and Bläser [24] in the period of 1980–1985 to grow crystals of more than thirty phases and determine their structures. The main difference from the scheme proposed by Zimmerman was the usage of a movable elliptic mirror to shift the focus point.

Further development of heating elements was related to the application of laser light sources. In 1992, scientific groups headed by Boese and Antipin [25,26] used an infrared laser in their devices for in situ crystal growth. This scheme was then commercialized and equipment for in situ crystallization began to be sold as the Optical Heating and Crystallization Device (OHCD) [27]. Presently, it is the only commercially available equipment for crystallization in thin-walled glass capillaries. Using this equipment, most of the structures published so far were crystallized from liquid samples and studied by X-ray diffraction. The experimental works related to in situ crystallization and subsequent X-ray diffraction studies allowed a big dataset to be collected. The quality of these data is increased, along with the progress in hardware and software used for X-ray diffraction. Early experiments in many cases resulted in very inaccurate structural data. There are several papers where information about the details of least-squares refinement (namely R-values, the values of s.u. for atomic coordinates) is missing. On the other hand, in some recent studies, the collected X-ray intensities were sufficient to carry out high-resolution experiments and multipolar refinement to simulate the charge density distribution. Thus, it is very hard to compare reliably, the bond lengths and angles from earlier and recent publications. For this reason, the main attention is paid to the analysis of crystal packing, especially short intermolecular contacts and packing motifs.

3. Organophosphorus, Organosilicon, and Organogermanium Compounds

In 1997, Karl Krueger's team published an article with the structures of several chelating organophosphines **1–5** (Scheme 1) [28]. In their article, they described the dependence of the P–C bond lengths on the organic substituent, showing a noticeable elongation of these bonds in cases where the steric hindrance of the substituent increased. In fact, for tris(tert-butyl)phosphine, the length of P–C bonds is 1.911(2) Å, and for tri(butyl)phosphine it is 1.844(2) Å [29]. Also, the sum of the C–P–C angles for all compounds is 303(3)°, with the exception of tris(tert-butyl)phosphine whose C–P–C angle is 322.3(3)°. To determine the crystal structure, the authors grew crystals of the compounds in situ in a diffractometer. A focused IR laser was used for zone melting. This was necessary because compounds **1** and **4** melt near 0 °C and the crystals that are formed are unstable, so it is difficult to transfer them to the diffractometer. Other compounds (**2** and **3**) melt at temperatures from −50 to −20 °C. Compound **5** melts above the room temperature, but the in situ methodology was used to obtain better single crystals too. A series of compounds studied in the article are organophosphines that contain two or four CH$_2$ groups between the phosphorus atoms. The authors explain that compounds with an odd number of methylene groups do not form crystalline phases upon cooling. In the crystal packing of all the compounds studied, the lone pairs of electrons of the phosphorus atoms look exactly in opposite directions, while in the case of three links in the linker they will probably look either in the same direction or orthogonally.

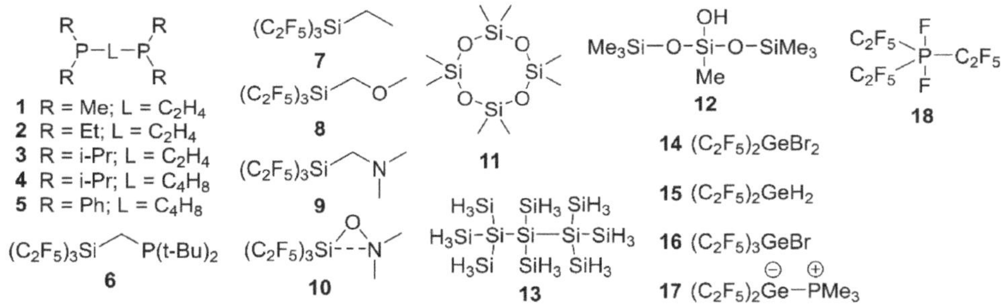

Scheme 1. Organophosphorus, organosilicon, and organogermanium compounds.

There are no strong intermolecular interactions in the crystal packing of compounds **1–5**. The molecules of these compounds are bonded by the H···H and C···H interactions. A dependency analysis between the calculated values of the molecular volume/surface area and the lattice energies indicates that the bonding between the molecules in the crystal of compound **5** is stronger than in the other ones. Conversely, in compounds **3** and **4**, the molecules are bonded weaker than it is predicted based on the hypothesis of a linear relationship of these quantities (Figure 1). It should be noted that the lone electron pair is not involved in any noticeable intermolecular interaction in crystals of **3** and **4**. On the contrary, in a crystal of compound **5**, the authors of [28], based on the analysis of the distribution of deformation electron density, established the interaction between the lone pair of electrons of phosphorus atoms and the phenyl group of the neighboring molecule. The authors of [28] believe that the above-mentioned interaction, along with the mutual steric influence of phenyl groups, is the result of the sp^3 hybridization of the phosphorus atoms. In turn, we can assume that it leads to a shrinkage of the crystal packing.

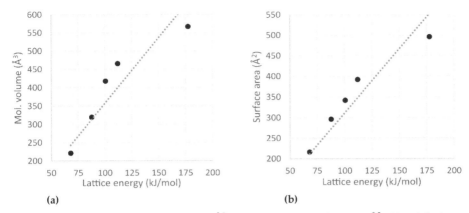

Figure 1. Correlation of molecular volume (Å³) (**a**) and Hirshfeld surface area (Å²) (**b**) with lattice energy (kJ/mol) of compounds **1–5**. Abscissa axis corresponds to lattice energy, while the ordinate axes correspond to molecular volume and surface area, respectively. Red dotted lines are trend lines.

In 2015 the phosphorous and silicon-containing compound $(C_2F_5)_3SiCH_2P(t\text{-}Bu)_2$ (compound **6**) containing a frustrated Lewis pair [30] was synthesized and its crystal structure was determined. The molecule is not involved in any noticeable intermolecular interaction, but the phosphorus atom is tetrahedral. After the addition of CO_2 or SO_2 to **6**, a heteroatomic ring is formed. The silicon atom acquires a distorted trigonal-bipyramidal environment because one of the oxygen atoms of the added molecule is bonded to this silicon. The phosphorus atom becomes bonded to a sulfur or carbon atom. Since compound **6** becomes a zwitterion after a gas is added, the intermolecular interaction energy grows and the temperature of melting goes above the room temperature.

In the crystals of CO_2 and SO_2 adducts, the C–Si–O angles between the opposite sides of bipyramids are 179.2° and 173.0°, respectively. The lengths of the Si–O bonds are 1.853 and 1.822 Å in the CO_2 and SO_2 adducts, respectively. The Si–C–P angles are 113.1° (CO_2 adduct), 117.1° (SO_2 adduct), and 120.4° (compound **6**). In all cases, these angles are bigger than the tetrahedral one (109.28°). Probably, the wide Si–C–P angle appears because of the high steric hindrance of silicon and phosphorus atoms, but this angle becomes smaller in adduct crystals because the five-membered ring "pulls together" the silicon and phosphorus atoms.

In the same year, a series of tris(pentafluoroethyl)silicon (TPFES) compounds was studied by Norbert W. Mitzel's team [31]. The purpose of that work was to search for intermolecular donor–acceptor interactions between the silicon atom and the electron pair of the β-atom in an α-TPFES-substituted compound (oxygen or nitrogen). The authors determined the structure of compounds with $-CH_2CH_3$ (**7**), $-CH_2OCH_3$ (**8**), $-CH_2N(CH_3)_2$ (**9**), and $-ON(CH_3)_2$ (**10**) groups. Based on interatomic distances, no interaction of the silicon atom with the β-atom was observed in the first three compounds. However, quantum chemistry calculations and X-ray structure analysis showed that this interaction appeared in the structure of compound **10** (Table 1).

Table 1. Comparison of experimental XRD data with the theoretical values (bond lengths are in Å, angles are in degrees) of some most important parameters of compounds **7–10** (data from a table in the original source [31]).

Compound	Parameter	XRD	B3LYP	PBE0	MP2
7	Si–C–C	118.6(2)	116.5	117.4	116.5
	Si···C	2.900(1)	2.910	2.907	2.893
8	Si–C–O	105.4(1)	106.1	107.1	104.3
	Si···O	2.619(1)	2.657	2.657	2.613
9	Si–C–N	115.5(2)	112.4	116.4	111.8
	Si···N	2.822(3)	2.803	2.847	2.777
10	Si–O–N	82.0(1)/83.7(1)	105.0	102.6	83.6
	Si···N	2.060(1)/2.093(1)	2.494	2.425	2.107

In order to establish whether the Si–N (Si–O) interaction was present, the X-ray diffraction experiment results were compared with the results of quantum chemistry calculations (B3LYP/6-31G(d,p), PBE0/cc–pVTZ, and MP2/cc–pVTZ). According to the authors, only the MP2/cc–pVTZ theory level allows one to approximate the experimental values cc-pVTZ of the Si···N distance for $(C_2F_5)_3SiONMe_2$. The experimentally obtained Si···N distance is smaller by 0.03 Å than that calculated using the MP2 method, and are 0.42 Å and 0.35 Å smaller, respectively, than those calculated by the B3LYP and PBE0 methods. If a methylene group between the silicon and nitrogen atoms existed, no Si-N interaction was found; therefore, the difference between the experiment and calculation was insignificant.

To explain the nature of the Si–N interaction, compound **10** was studied by gas electron diffraction. These molecules are conformationally flexible and the rotation barrier around every single bond is small. To evaluate the conformational flexibility of compound **10** in the gas phase, the experimental radial distribution function of interatomic distances was compared with the distribution of interatomic distances obtained from the calculation by the molecular dynamics method. The R-factors for the radial distribution found by molecular dynamics do not exceed 5.2%. The conformation whose radial distribution function is reproduced most accurately (4.0%) does not contain a short Si···N contact. Thus, this contact is found only in a crystal environment and is probably caused by steric hindrance.

The crystal packing of compounds **7–10** is stabilized mainly due to weak H···F hydrogen bonds and F···F interactions. The F···F interatomic distances of intermolecular interactions are only slightly (by no more than 0.15 Å) shorter than the sum of the corresponding van der Waals radii. The only exception is the structure of compound **10**, where the F···F interactions are stronger: some F···F distances are shorter than the sum of the van der Waals radii by more than 0.2 Å. However, the F···F interactions become stronger because of the disordering of one of the pentafluoroethyl groups. This is consistent with the fact that the crystal lattice energy (Table S1 in Supplementary Materials) of compounds **7–9** varies in a very narrow range and is almost independent of the composition.

In 1955, the structure of the low-temperature phase of octamethylcyclotetrasiloxane (compound **11**) was established [32]. This compound crystallizes at 17.5 °C and has a phase transition at −16.3 °C [33,34]. The unit cell of the low-temperature phase is tetragonal, belongs to the $P4_2/n$ space group, and its parameters are $a = b = 16.10 \pm 0.02$ Å, $c = 6.47 \pm 0.01$ Å [32]. The molecules are in the pseudo-chair conformation in the low-temperature phase. For the high-temperature phase, only the space group ($I4_1/n$) and the cell parameters were determined. The authors report that the unit cell parameters a and b change insignificantly ($\pm$ 0.02 Å) during the phase transition, while the parameter c changes from 6.47 Å to 6.83 Å.

In 2018, we published an article about the synthesis of siloxanols [35]. We determined the structure of 1,1,1,3,5,5,5-heptamethyltrisiloxan-2-ol (compound **12**) that is viscous at room temperature. It forms a strongly bonded (with O–H···O hydrogen bond) tetramer around the $\bar{4}$ axes in the crystal (Figure 2). The O···O distance in the hydrogen bond is as short as 2.711(4) Å, which means that O–H···O is a strong hydrogen bond.

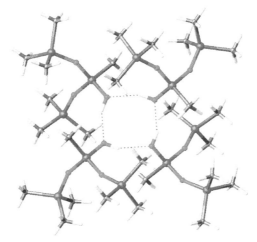

Figure 2. Tetramer in a 1,1,1,3,5,5,5-heptamethyltrisiloxan-2-ol crystal (compound **12**).

In 2017, Harald Stueger's team obtained and studied 2,2,3,3,4,4-hexasilylpentasilane (compound **13**) [36]. This compound is a liquid, and the in situ crystallization method was used to determine its crystal structure. The synthetic method is also applicable for the subsequent synthesis of oligomers and polymers with similar structures.

In 2016, Berthold Hoge's team published a series of works devoted to the synthesis and properties of perfluoroethyl-substituted organogermanium compounds [37–39]. In these works, some physicochemical properties of compounds were examined, such as vibrations in the IR spectrum and crystal packing. Four of the eight compounds, namely, $(C_2F_5)_2GeBr_2$ (**14**), $(C_2F_5)_2GeH_2$ (**15**), $(C_2F_5)_3GeBr$ (**16**), and $(C_2F_5)_2GePMe_3$ (**17**), are liquids at room temperature. In order to determine the structures of their crystalline phases by X-ray diffraction experiment, these samples were crystallized in situ.

A further study of the crystal structures of all four compounds showed that compounds **14** and **16** have similar intermolecular contacts but different packing motives. In the crystal, molecules **14** form stacks in a square packing (Figure 3). The crystals of **15** and **16** form layered structures. The propensity of perfluorinated groups to form layered structures is also shown in [40]. The layers in the crystal of compound **15** are identical and are composed of molecules in an all-*trans* conformation. The interactions between the layers in the crystal of **16** alternate: half pairs of layers are connected only by F···F contacts, while the other pairs of layers are mainly connected by Br···F interactions (Figure 3, right).

| 14 | 15 | 16 |

Figure 3. Crystal packing motifs of $(C_2F_5)_2GeBr_2$ (**14**), $(C_2F_5)_2GeH_2$ (**15**), and $(C_2F_5)_3GeBr$ (**16**).

The germanium atom in compound **17** is formally divalent and has a lone electron pair. As a result, the lone electron pair of the germanium atom is involved in the C–H⋯Ge intermolecular interaction with the methyl group at the phosphorus atom. The C⋯Ge distance is 3.1793(6) Å, which is shorter than the sum of the van der Waals radii (3.31 Å). The Ge–P bond length is only 2.3989 (16) Å, which is only 0.14 Å longer than the sum of the covalent radii (2.27 Å). Otherwise, the crystal of this compound is not very different from the structure of compound **15**: H⋯F and F⋯F contacts are also present.

Another interesting result of Hoge's team was published in the European Journal of Inorganic Chemistry [41]. This work was devoted to the study of various trisubstituted phosphines, di and tri-fluorophosphates. Compounds with acceptor substituents such as pentafluoroethyl, pentafluorophenyl, and tetrafluoropyridyl were prepared.

The most interesting example is a crystal of the compound $(C_2F_5)_3PF_2$ (**18**), in which the molecules are bonded only via F⋯F contacts. In this case, the fluorine atoms of the PF_2-group are not involved in noticeable intermolecular interactions. The shortest distances F⋯F involving PF_2 groups are 2.995(7) Å (the sum of van der Waals radii equals 3.31 Å).

4. Organosulfur Compounds

In 1998, Yoshihiro Yokoyama and Yuji Ohashi published an article in which they described the crystal structures of 1-methoxy-2-(methylthio)ethane (MMTE, compound **19**) and 1,2-bis(methylthio)ethane (BMTE, compound **20**) (Scheme 2) [42]. Crystals of these compounds were grown by in situ crystallization at a temperature of 10 degrees below the melting point. In this case, recrystallization was carried out by the partial melting of a polycrystalline sample. The authors note that the main problem of this method of crystal growth is the chance of accidentally melting the entire sample.

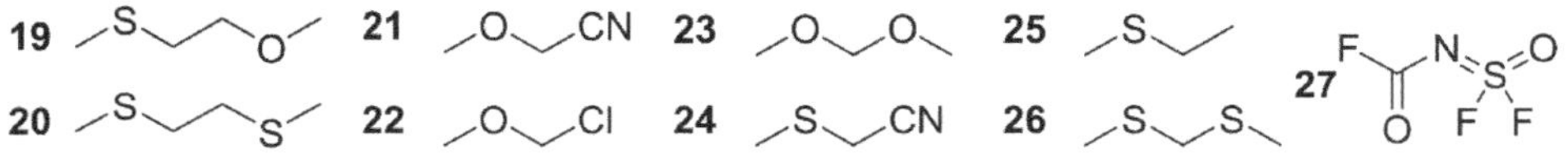

Scheme 2. Organosulfur and oxygen-containing compounds.

The molecules in the crystals of compounds **19** and **20** are bonded mainly due to the weak dipole–dipole interactions of methyl hydrogen atoms with lone electron pairs of sulfur or oxygen atoms. The crystals belong to the $P2_1/n$ (for MMTE **19**) and $P2_1/c$ (for BMTE **20**) space groups. The authors note that in both crystal packings, the molecules are in almost identical conformations. Indeed, the S–C–C–S and O–C–C–S torsion angles are 180 and 178°, and the C–S–C–C angles are 71 and 79° for compounds **19** and **20**, respectively.

To determine the relative stability of possible conformations, quantum chemical calculations were performed. According to these calculations, the SC–CS-*trans*-conformer is more stable than the *gauche*-conformer, while the CS–CC-*gauche*-conformer is more stable than the *trans*-conformer. Based on the melting points of these compounds and the melting point of 1,2-dimethoxyethane (DME), the authors suggest that the BMTE crystal, which has the highest melting point, is the most stable among them. This was also confirmed by quantum chemical calculations and experiments on crystal growth from binary mixtures of these compounds. Crystals of **20** were grown from the BMTE:MMTE = 1:1 and BMTE:DME = 1:1 mixtures.

In 1999, the same authors [42] published an article where they compared the crystal structures of compounds with general formula XCH_2OCH_3 (X = CN (**21**), Cl (**22**), OCH_3 (**23**)) and XCH_2SCH_3 (X = CN (**24**), CH_3 (**25**), OCH_3 (**26**)) [43] obtained by the same method as in [42]. According to the results of quantum-chemical calculations (MP2/6-31G*), the *gauche* conformation is more favorable than the *trans* conformation. This is probably due to the presence of a strong anomeric effect. In this case, the *gauche* conformation realized in the crystal becomes the most beneficial for an isolated molecule.

The difluorosulfenylamide cyanide is a conformationally flexible compound due to the presence of hindered rotation around the formally double S=N bond. To investigate the conformational flexibility of this compound, Cutín and colleagues [44] carried out a gas electron diffraction experiment. Later, Roland Boese et al. [45] used X-ray diffraction with the in situ crystallization method to determine the structure of the related solid-phase compound ((fluoroformyl)imidosulfuryl difluoride, compound **27**). It revealed that compound **27** is in the antiperiplanar-synperiplanar conformation. To reveal the relative stability of the possible conformations, quantum chemical calculations (HF/6-31G+*, B3LYP/6-31G+*, B3LYP / 6-311G+*, and MP2/6-311G+*) were performed. The lowest-energy conformation is antiperiplanar-synperiplanar and constitutes 69–80% of the molecules at room temperature (the assessment is based on the Boltzmann distribution). The fraction of the synclinal-synperiplanar conformation is 12–23% of the molecules, while the fraction of the others is small.

5. Organohalogen Compounds

The team of Roland Boese, Ashwini Nangia, and Gautam R. Desiraju described the intermolecular interactions in the crystals of partially fluorinated benzenes **28–34** (Scheme 3) [46]. The subject of the study included mono, *ortho*, and *para*-bi-, tri-, tetra-, and penta-fluorobenzenes, and *para*-halo-substituted fluorobenzenes. Since all these compounds have low-melting points from 225 to 277 K, the in situ crystallization method was used for the crystal growth. An IR laser was used for zone melting.

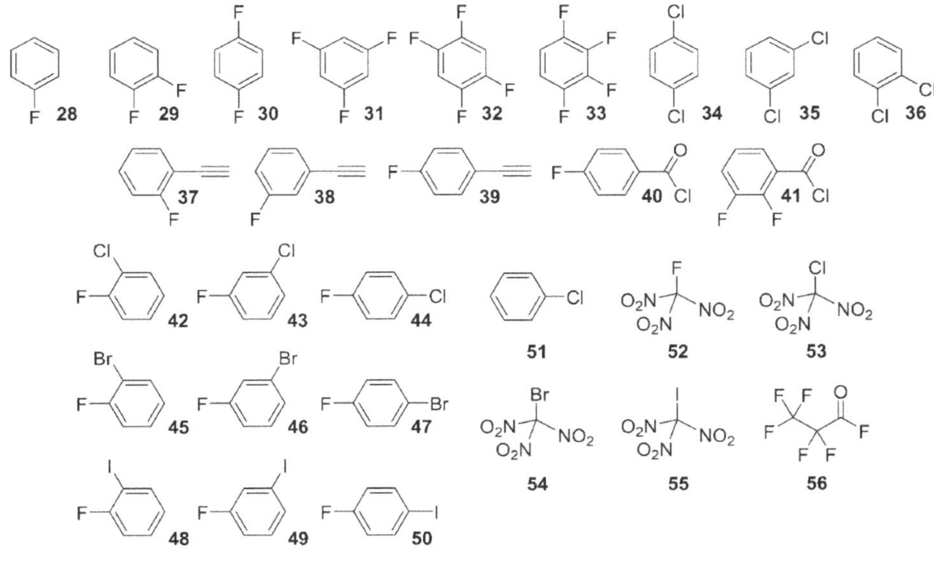

Scheme 3. Halogen-containing compounds.

In a monofluorobenzene **28** crystal, the main type of interatomic interaction is the weak C–H···F hydrogen bond. At the same time, the presence of the C–H···π and F···F interactions was also observed. According to the authors, the presence of these interactions makes the crystal packing of fluorobenzene similar to that in Py·HF, C_5H_5NO, and PhCN [46]. It is interesting to note that the crystal packings of fluorobenzene and chlorobenzene are significantly different, while the crystals of fluorobenzene and benzonitrile are isostructural (if the fluoro and cyano groups are not taken into account). In the structures of difluorobenzenes, the molecules are packed in layers. The molecules in the layers of *ortho*-difluorobenzene **29** interact with each other due to weak C–H···F hydrogen bonds and stacking interactions. At the same time, the interaction between the layers is due to the interaction of the lone pairs of fluorine atoms with the π-system of benzene rings (F$\cdots\pi$). On the contrary, in a *para*-difluorobenzene **30** crystal, layers are formed only due to weak hydrogen bonds. No interactions involving the π-system of the benzene ring were observed. In turn, the layers bind to each other through stacking and dipole-dipole interactions of the phenyl rings. The authors note explicit relations between the crystal packing of 1,4-benzoquinone and *para*-difluorobenzene, which indicates that these molecules have similar electronic structures [46]. Moreover, the crystals of these compounds are isostructural (the cell parameters and space groups coincide).

Similarly, the structures of dihalobenzene crystals are also similar to each other. In all these compounds (except the previously described *para*-difluorobenzene), pronounced halogen bonds are present (Cl$\cdots$Cl, Br$\cdots$Br, F$\cdots$I, all belonging to the second type [47]) and the orientations of molecules in the layers alternate (Figure 4).

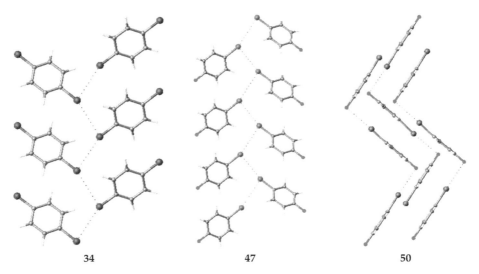

Figure 4. Crystal packing motifs of *p*-dichlorobenzene **34**, *p*-bromofluorobenzene **47**, and *p*-iodofluorobenzene **50**. The halogen bonds close to the L type ($\theta_1 > 155°$, $\theta_2 \approx 90°$) are shown as dotted lines.

The crystal packing of trifluorobenzene **31** is pseudohexagonal. The authors of [46] draw a clear analogy with the packaging of 1,3,5-triazine [48,49]. In both cases, there is an electrophilic hydrogen atom interacting with two electron pairs of neighboring molecules, which allows one to predict the "hexagonal cell" structure.

The crystal packing of tetra-fluorobenzenes varies greatly. The crystal structure of 1,2,4,5-tetra-fluorobenzene **32** [46] is similar to that of tetrazine [50]. In these structures, molecules in one layer are bonded through dipole-dipole interactions between hydrogen atoms and lone pairs of a neighboring molecule. 1,2,3,4-Tetra-fluorobenzene **33** crystallizes in two polymorphic modifications. The first polymorph was grown at a temperature of 123 K close to the melting point, while the second polymorph was crystallized at a temperature of 195 K from the toluene:pentane system (1:3). Interactions of C–H$\cdots$F type contribute a lot to the stabilization energy of both polymorphs. In the first polymorph, the molecules form stacking interactions between the layers. In the second one, another motive exists due to the bonding of this type: pairs of molecules are rotated relative to each other by an angle of almost 90° and alternate in a checkerboard pattern. A similar packing is also observed in a pentafluorobenzene crystal. However, dipole–dipole interactions are weaker due to a decrease in the number of hydrogen atoms.

Based on the studies, the authors of [46] concluded that crystalline packing is determined by the presence of acceptor fluorine atoms. The type of packaging depends on the number of these atoms.

Para-dichlorobenzene **33** exists in the form of three polymorphs: α (**34a**), β (**34b**), and γ (**34c**). According to Wheeler and Colson, the number of shortened Cl···Cl contacts (<3.9 Å) increases from three in the β polymorph to four in the α polymorph and five in the γ polymorph, which corresponds to the sequence of phase transition with decreasing temperature [51,52]. A CE–HF/3-21G calculation of the packing energy performed by us predicts that the γ-polymorph is the most stable. In this case, the energies of α and β-polymorphs differ insignificantly (by 1.7 kJ/mol), while the latter is more stable. In 2001, Roland Boese et al. grew crystals of dichlorobenzenes **35** and **36** by the in situ method [53]. To predict the stability of the crystalline structures studied, calculations that included a computer search for the most thermodynamically stable modifications (UNI force field [54]) were performed. The structures generated in the computer search were compared with the results of X-ray diffraction experiments carried out at temperatures of 100 and 220 K for *ortho* (**36**) and *meta*-dichlorobenzenes (**35**). The parameters of the most thermodynamically stable crystal cells obtained from the calculation are generally consistent with the experimental ones for all the phases. The only exception is the β-polymorph of *para*-dichlorobenzene (**34b**), for which the predicted cell parameters differ noticeably from the experimental ones.

In 2011, Desiraju et al. compared the crystal structures of phenylacetylene and its monofluorinated derivatives [55]. For each of the 2- and 3-fluoro-substituted phenylacetylene derivatives (**37** and **38**, respectively), the existence of two crystalline phases was determined. The orthorhombic phase **37a** (Form 1) crystallizes in the orthorhombic cell (space group Pna2$_1$). The molecules in this phase are disordered in such a way that the ratio of the population of fluorine atoms on one or the other side of the acetylene fragment is approximately 1:4. In this phase, there is an interaction of the terminal hydrogen atom of the acetylene fragment with the π-system of the acetylene fragment of the neighboring molecule, and non-classical hydrogen bonds of the C–H···F type. Due to these interactions, a zigzag motif (Figure 5) of molecules along the *c* axis is present in the crystalline packing. Two types of weak C–H···F hydrogen bonds were found in polymorph **37a**: in the first case, the hydrogen atom of the acetylene fragment participates in the bonding, and in the second one, the phenyl ring hydrogen atom interacts with the fluorine atom. Polymorph **37b**, which was obtained by slow cooling of a liquid in a capillary, crystallizes in space group P2$_1$. In the crystal packing of this polymorph, the interaction of the terminal hydrogen atom of the acetylene fragment with the π-system of the acetylene fragment of the neighboring molecule also exists, but unlike **37a**, only the hydrogen atoms of the phenyl fragment participate in the C–H···F weak hydrogen bond.

Both polymorphs of (3-fluorophenyl)acetylene **38** (forms 1 (**38a**) and 2 (**38b**)) are monoclinic, but they are characterized by noticeably different cell volumes and different space groups. In fact, **38a** crystallizes in the centrosymmetric group P2$_1$/n and contains three molecules in the unique part of the cell. The molecules are notable for a hydrogen atom in the phenyl ring which forms the strongest hydrogen bond. It should be noted that the hydrogen atom of the acetylene fragment forms a C–H···F hydrogen bond only in one molecule. In the other two molecules, the same terminal hydrogen atom interacts with the π-system of the acetylene fragment of another molecule. In the **38b** crystal, like the **37b** crystal, a zigzag-like interaction is present (Figure 6).

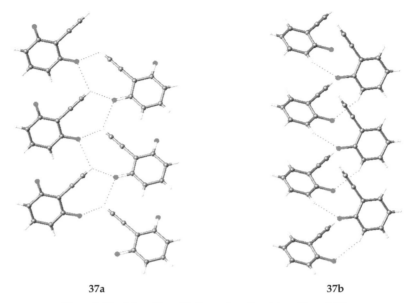

37a   37b

Figure 5. Crystal packing of (*o*-fluorophenyl)acetylene **37a** and **37b**.

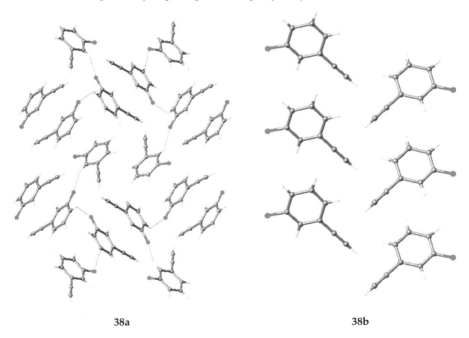

38a   38b

Figure 6. Crystal packing of (*m*-fluorophenyl)acetylene **38a** and **38b**.

(*Para*-fluorophenyl)acetylene **39** consists of layers in which the molecules are bonded by hydrogen bonds between the terminal acetylene fragment and the fluorine atom of the neighboring molecule in one direction and by stacking in the other direction. The layers are interconnected by weak H···H contacts. As a result, for *ortho* and *meta*-substituted monofluorophenylacetylenes, the main motive of crystal packing is the zigzag-like interaction of acetylene fragments, and for *para*-substituted ones, it is stacking.

It is known that interactions between halogens are usually formed due to the presence of a σ-hole of one atom, which interacts with the electron-saturated "belt" of another atom [56]. Desiraju shows the interactions of cis-trans geometry and L-geometry [47]. The most remarkable examples of compounds that can have crystals with halogen bonds include 4-fluorobenzoyl chloride **40** and 2,3-difluorobenzoyl chloride **41** [57]. The Cl···F interactions in these crystals belong to the second type of geometry (L-geometry). It should be noted that different atoms act as donors and acceptors of a halogen bond. In a crystal of **40a**, the fluorine atom is the halogen bond donor and the chlorine atom is the acceptor, while in a crystal of **41**, the chlorine atom is the donor and the *ortho*-fluorine atom is the acceptor. In both crystals, oxygen is involved in the formation of two hydrogen bonds with *ortho* and *meta*, or *meta* and *para* hydrogens. The second fluorine atom (*meta*-fluorine) in compound **41** forms a hydrogen bond with the *meta* hydrogen atom of a neighboring molecule, and in compound **40a** a hydrogen bond is formed by an *ortho*-atom with the single fluorine atom. Thus, the authors concluded that the chlorine atom of the acyl chloride group can act both as a donor and as an acceptor.

Two years later, the same authors published an article in which they described the intermediate phase **40b** formed during the crystallization of **40** [58]. The latter crystallized in the same space group and was characterized by a similar layered type of packing. In that work, the authors compared the molecular conformations in crystal phases with the ones calculated for the gas phase. The locations of the molecules within the layers are almost the same for both polymorphs. However, in the intermediate phase, the rings of the molecules form a parquet type packing, while in the previously studied phase **40a**, the phenyl rings of the molecules are parallel to the planes of the layers. In addition, the layers are shifted relative to each other. In the layers of both phases, there are significant differences in the length of the Cl···F halogen bond (3.153 Å for the form **40a** and 3.283 Å for the form **40b**). This is due to the presence of a staircase structure. According to the authors, the crystals of the phase **40b** are gradually transformed to phase **40a**. Moreover, a crystal can be represented as a combination of domains of both phases during the experiment. Our calculations (Table S1) are in agreement with the conclusion about the instability of **40b** (**40a** is about 5 kJ/mol more favorable than **40b**).

Hirshfeld surface analysis (carried out using CrystalExplorer, Version 3.1, [9,59]) was performed by the authors for both crystal structures, and the diagrams of fingerprints of intermolecular interaction were calculated. The main contribution to the Hirschfeld surface corresponds to the C–H···O, C–H···F, and C–H···Cl hydrogen bonds. Nevertheless, the aromatic cycle stacking also makes a significant contribution to the **40b** phase surface, while in the stable phase **40a**, π–π interactions between the phenyl substituent and the anhydride group are observed instead. The C–C–C–Cl torsion angle between these fragments of the molecule in the **40b** phase is as small as 0.4(8)°, while in the crystal of phase **40a** it is 11.6(2)°. The deviation from the plane conformation in a crystal of **40a** is probably caused by the formation of a stronger F···Cl halogen bond.

Dikundwar et al. published an article comparing the chloro, bromo, and iodo-derivatives of fluorobenzene [60]. The main goal of that work was to determine the influence of the type of halogen and its position on the formation and geometry of the halogen bond. To determine the crystal structure of the compounds, nine crystals were grown by the in situ method. Two phases were found for *meta*-chlorofluorobenzene, but only one for each of the remaining compounds. *Ortho*-chlorofluorobenzene **42** molecules do not form halogen bonds; instead, they participate in the formation of hydrogen bonds and C–H···π interactions. In the first phase of *meta*-chlorofluorobenzene **43a**, the Cl···Cl halogen bonds exist in the L geometry, due to which the molecules form zigzag-like motifs in the crystal packing. In the other phase **43b**, halogen bonds are not formed and all intermolecular interactions are related to hydrogen bonds. In a crystal of *para*-chlorofluorobenzene **44**, which was first described by Boese and Desiraju [46] and later studied by Sarah Masters' team [61], zigzag-like halogen bonds are observed, like in the first phase of *meta*-chlorofluorobenzene **43a**, but they already correspond to the trans-type. *Ortho*-bromofluorobenzene **45** forms zigzag-like chains in which molecules are bonded by Br···π interactions. In addition, in a **45** crystal, halogen bonds with both fluorine and bromine exist. A *meta*-bromofluorobenzene **46** crystal contains two independent molecules that form the only trans-type halogen bond (Br···Br). The structure of *para*-bromofluorobenzene **47** was previously determined by Boese and Desiraju [46]. In a crystal of this compound, it was found that the bromine atoms interact with each other through first-type halogen contacts, while the fluorine atoms participate only in the formation of hydrogen bonds. A similar situation is observed in a crystal of *ortho*-iodofluorobenzene **48**, but, unlike compound **47**, the I···I halogen bonds form a zigzag-like motif and belong to the second type. The structure of *meta*-iodofluorobenzene **49** differs significantly from that described previously. This compound crystallizes in space group $P2_1$, with five molecules in the unique part. Three of them are arranged in spirals along the screw pseudo-axis 3_1 (Figure 7). The molecules around the pseudo-axis form second-type halogen bonds with each other ($\theta_1 \approx 180°$, $\theta_2 \approx 90°$). The remaining two molecules form zigzag-like structures based on halogen contacts, but closer to the first type ($\theta_1 \approx 155°$, $\theta_2 \approx 124°$, Figure 7). The structure of *para*-iodofluorobenzene **50**, also previously studied by the Boese and Desiraju teams [46], contains a zigzag-like structure of the I···I halogen bond, but F···I halogen bonds also exist.

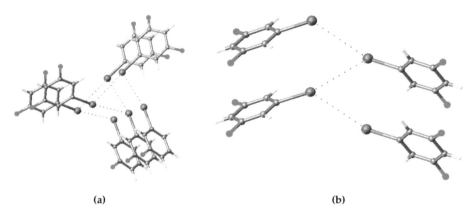

(a)   (b)

Figure 7. Crystal packing motifs in a crystal of *meta*-fluoroiodobenzene **49**: molecules around pseudo-axis 3_1 (**a**) and axis 2_1 (**b**).

Based on the above, fluorine atoms are more likely to form hydrogen bonds than halogen bonds, while heavier halogens behave in the opposite manner. According to the authors, this is due to the principle of "like likes like", since the sizes of the fluorine and hydrogen atoms are similar.

In 2015, Nath and Naumov published the structure of a crystal of chlorobenzene **51** [62]. To analyze the intermolecular interactions, the analysis of Hirschfeld partitioning was carried out for chlorobenzene. The calculation showed that a significant fraction of the surface (32.2%) corresponds to the π(C)···H contacts. The H···H and H···Cl contacts contribute 35.1% and 25.9% to the Hirschfeld surface, respectively. Also, in a crystal of chlorobenzene, the zigzag-like motif consisting of molecules bound by a halogen bond exists.

The Pierangelo Metrangolo and Giuseppe Resnati teams determined the crystal structure of a number of co-crystals of amines with halo-pentafluorobenzenes in which a halogen bond is present [63]. This type of intermolecular bond is formed between bromine or iodine atoms in C_6F_5I or C_6F_5Br, respectively, and a lone pair of electrons of the nitrogen atom of substituted pyridine or TMEDA. The I···N distances in the compounds studied are 2.784 Å on average, which is shorter than the Br···N distance found (2.882 Å). Nevertheless, all these distances are much shorter than the corresponding sums of the van der Waals radii (3.53 Å for I⋅⋅⋅N and 3.40 Å for Br⋅⋅⋅N), which definitely indicates that a halogen bond exists in each of the crystals studied.

In 2014, Klapötke et al. published an article in which they compared the structures of halo-trinitromethanes **52–55** obtained from gas electron diffraction, X-ray diffraction analysis, and quantum chemical calculations [64]. Fluoro and bromo-trinitromethane crystals were grown in situ in a capillary using copper wire as the heating element. According to the authors, contrary to the literature, iodotrinitromethane did not decompose under MoKα radiation, as was stated in [65]. The X-ray diffraction data for chlorotrinitromethane was taken from an earlier article [66]. The results of X-ray diffraction experiments were compared with those from gas electron diffraction. The relative arrangement of nitro groups in the crystal and in the gas phase can be described as a "propeller". The lengths of C–Hal bonds in a crystal are shorter than in the gas phase, while the opposite pattern is observed for C–N bonds. From the analysis of structural data, it was concluded that a difference in the type of halogen atoms has almost no effect on the C–N bond lengths and N–C–Hal angles in these compounds. On the other hand, the torsion angle of the nitro group has a tendency to increase with an increase in the atomic number of the halogen. As the van der Waals radius of the halogen atoms grows, the intra- and intermolecular interactions between them and the oxygen atoms of the nitro groups increase. According to the authors, intramolecular O⋅⋅⋅Hal interactions are forced, while intermolecular interactions are advantageous and can be considered as analogs of the L-type halogen bond. Using the NBO, AIM, and IQA methods, the atomic charges, atomic volumes, and interatomic interaction energies were calculated. IQA was performed using the RHF/cc–pVTZ wavefunctions for fluoro and chloro-derivatives and RHF/6-311G(d) for bromo and iodo-derivatives. Both NBO and AIM agree that a negative charge is present on fluorine atom, while it is positive on the other halogens. At the same time, the positive charge on carbon atoms decreases as the halogen atom increases. The charges on nitrogen and oxygen atoms do not depend on the type of halogen at all.

In 2016, Norbert W. Mitzel and Carlos O. Della Védova's teams published a joint work in which perfluoropropanoic acid fluoride **56** (CF$_3$CF$_2$C(O)F) was described [67]. The structure of this compound was studied by gas electron diffraction, IR, Raman, and ultraviolet spectroscopy, and by quantum chemical calculations (MP2/cc–pVTZ and B3LYP/cc–pVTZ). Compound **56** is a volatile liquid at room temperature, with a melting point of about 146 K. In order to determine the crystal structure, X-ray diffraction analysis and in situ crystallization were used. The crystal was grown at 144 K, after which the sample was slowly cooled to 100 K for the X-ray diffraction experiment. The results showed that all molecules in the crystal are in gauche-conformation, while in the gas phase, according to the results of gas electron diffraction, an equilibrium exists. It was shown that only 85(10)% of molecules are in the gauche-conformation, while the remaining 15(10)% are in the anti-conformation.

6. Intermolecular Interactions in the In Situ Crystallized Compounds

Analysis of crystal packing allowed us to conclude that organoelement compounds described in the present review can be divided into two groups. The first group consists of Si, Ge, P, and S-containing compounds. The heteroatoms in compounds of the first group are surrounded with a shell of hydrocarbon substituents (including perfluorinated ones) and do not participate in any intermolecular interactions. As a consequence, all structure-forming intermolecular interactions are weak classic and non-classic hydrogen bonds (X⋅⋅⋅H, H⋅⋅⋅H, C⋅⋅⋅H; X = Hal, O, N, C). The second group includes organofluorine aromatic compounds where intermolecular interactions between fluorine atoms play a significant role in the stabilization of crystal packing. Moreover, fluorine atoms' interactions can be of many types—weak hydrogen bonds, halogen bonds, F⋅⋅⋅F interactions, and F⋅⋅⋅π interactions [68]. Besides, a noticeable contribution to the stabilization of crystal packing is played by other intermolecular interactions with the participation of the π-systems of phenyl rings [69,70].

According to the calculation of the energy frameworks, compounds **1–17** can be described as loosely packed ones. The ratio between the molecular volume (Å^3) and the lattice energy (kJ/mol) in the majority of these compounds is about 3.5-5 Å^3·mol/kJ (Table S1). This fact can be explained by the dominance of H···H and C–H···π interactions over others (except for the strong hydrogen bond present in a crystal of **12** compound). The situation is quite different in the case of organosulfur compounds **21–26** where the ratio varies within 2–3. The reason lies in the large contribution of weak C–H···O or C–H···N hydrogen bonds that exceed the contribution from the C···H, H···H, and H···π bonds. Possibly, the correlation between molecular volume and lattice energy can be explained by strong F···F interactions. Organofluorine aromatic compounds have the ratio value close to three due to a contribution from interactions of fluorine and other halogens.

The correlations of the molecular volume/surface area with the lattice energy for all the compounds discussed above are shown in Figure 8. Most of the compounds are liquids at room temperature and have molecular volumes in a range of 120–160 Å^3. The Hirshfeld surface areas of the majority of compounds do not exceed 170 Å^2. The maximum lattice energy of compounds that are liquid at room temperature is 112 kJ/mol (compound **4**). However, most of the compounds have lattice energies between 45 and 65 kJ/mol.

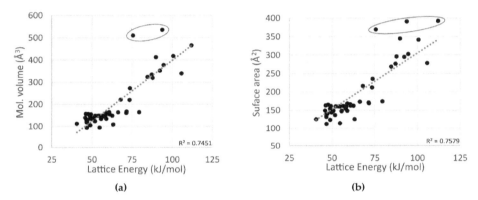

Figure 8. Correlation of molecular volume (Å^3) (**a**) and Hirshfeld surface area (Å^2) (**b**) with lattice energy (kJ/mol) for the majority of compounds (the full list of compounds is shown in Table S2). Red dotted lines are linear trend lines.

The trend line was calculated on the assumption of a linear relationship between the molecular surface area, volume, and lattice energy values. The R^2 values for these trend lines are 0.7451 for plot (**a**) and 0.7579 for plot (**b**), respectively (Figure 8). The compounds that are solid at room temperature and the phases that contain strongly disordered molecules were excluded from these relationships.

Several points in Figure 8 lie far from the trend line (the points encircled in red lines). These are compounds **4, 6, 13** (compound **4** is encircled only in part (**b**)). These compounds stand out because most of the intermolecular interactions in their crystals are weak, but the molecules are still big (the molar volumes are above 400 Å^3).

Relatively low R^2 values for the trend lines in Figure 8 indicate a poor approximation of the entire data set. In this case, we separated the compounds into three groups. The separation is based on the largest Hirshfeld surface area occupied by the intermolecular interaction of a certain type (Figure 9, Tables S2 and S3). Compounds were separated into "H···H", "H···Hal" (Hal = F, Cl, Br, I), and "Other" groups.

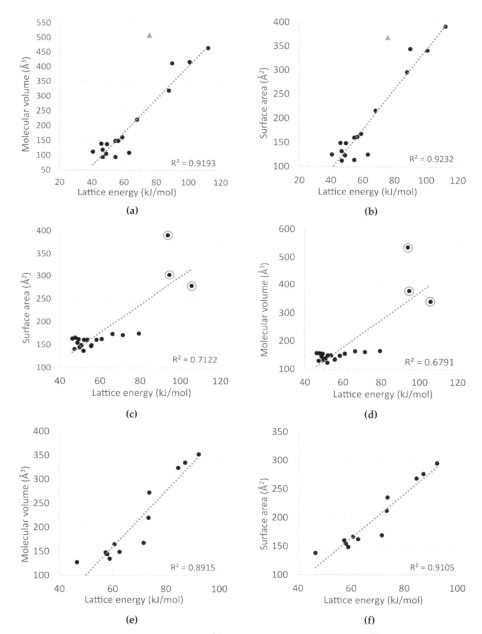

Figure 9. Correlation of molecular volume (Å^3) (**a,c,e**) and Hirshfeld surface area (Å^2) (**b,d,f**) with lattice energy (kJ/mol) for the compounds with predominant H···H interactions (**a,b**), Hal···H interactions (**c,d**), and other ones (**e,f**). Red triangle (**a,b**) corresponds to 2,2,3,3,4,4-hexasilylpentasilane (compound **13**). Three rounded dots (**c,d**) correspond to organophosphorus compounds **6**, **9**, and **17**. Red dotted lines are linear trend lines.

The first "H⋯H" group is well approximated linearly (Figure 9a,b). One point (red triangle) was excluded. It corresponds to compound **13**. In this crystal, Si–H⋯H–Si intermolecular contacts are observed instead of C–H⋯H–C contacts. The Si–H⋯H–Si interaction is much weaker, and the lattice energy is at least 30 kJ/mol smaller than that predicted by the trend line. The second group "H⋯Hal" is approximated much worse (Figure 9c,d). Most of the compounds are small molecules. Their Hirshfeld surface area and molecular volume do not exceed 200 Å²/Å³. Three rounded dots correspond to organophosphorus compounds **6, 9**, and **17**. The last group, "Others", is well approximated linearly, like the first one (Figure 9e,f).

The dependence of the melting point on the lattice energy is shown in Figure 10. We separated all data into black, blue, and red groups. Black dots correspond to compounds with a molecular surface area below 170 Å². Blue triangles correspond to compounds whose molecular surface areas are above 170 Å². The red square corresponds to compound **1**. The tangent value (m.p./lattice energy) for the "black" group is between 3.13 and 6.10, and 4.75 on average. The same value for the "blue" group is between 1.63 and 3.12, and 2.41 on average. The molecules of compounds of the "black" group are small and most of them do not form strong intermolecular contacts. On the other hand, in the crystals of the "blue" group compounds, molecular interactions such as halogen bonds or C–H⋯P hydrogen bonds are present. Compound **1** was assigned to a separate group because the C–H⋯P bond existing in its crystal is not strong enough, but the Hirshfeld surface area is too large for the "black" group. It means that this compound should be in the "black" group because of its tangent (m.p./lattice energy) value, but it has potentially strong C–H⋯P interactions.

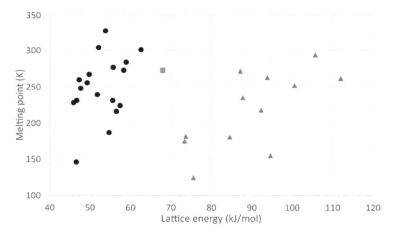

Figure 10. The plot of melting point (K) versus lattice energy (kJ/mol). Black dots correspond to compounds with a molecular surface area below 170 Å². Blue triangles correspond to compounds whose molecular surface area is above 170 Å². The red square corresponds to compound **1**.

7. Conclusions

Organoelement compounds, which melt below room temperature, usually have not got any strong intermolecular interactions in their crystals. The significant role in crystal packing is played by medium strength interactions, such as halogen bonds. In this review, the structures of 56 in situ crystallized compounds were discussed. In some crystals, such as 1,1,1,3,5,5,5-heptamethyltrisiloxan-2-ol crystal (compound **12**), we see the presence of a strong interaction. Even in these crystals, the weak intermolecular interactions prevail. The physical properties of all in situ crystallized compounds could not be well approximated. But compounds with similar structures have similar interactions in their crystals. So, the compounds were divided into groups by predominant interactions. After the division into groups, the correlation of molecular volume ($Å^3$) and Hirshfeld surface area ($Å^2$) with lattice energy (kJ/mol) became much clearer.

Supplementary Materials: The following are available online at http://www.mdpi.com/2073-4352/10/1/15/s1: Quantum chemistry calculation details, Table S1. Sublimation energies, melting points, molecular volume, and surface area of the compounds, Table S2. Intermolecular interactions in the compounds, Table S3. Intermolecular interactions in halogentrinitromethanes 52–55.

Author Contributions: The sections "In Situ Crystallization: A Retrospective View" and "Organophosphorus, Organosilicon, and Organogermanium Compounds" were written by A.D.V. The section "Organohalogen Compounds" was written by A.F.S. The sections "Introduction" and "Organosulfur compounds" were written by A.A.K. The section "Summary" was written by all the authors. All authors have read and agreed to the published version of the manuscript.

Funding: This work was financially supported by the Russian Science Foundation (project 18-73-00257), the Russian Foundation for Basic Research (project 19-33-90196), and the Ministry of Science and Higher Education of the Russian Federation.

Acknowledgments: The quantum chemistry calculations were performed with the financial support of the Russian Science Foundation (project 18-73-00257). The study of crystal packing was performed with the financial support of the Russian Foundation for Basic Research (project 19-33-90196). The analysis of organofluorine compounds was supported by the Ministry of Science and Higher Education of the Russian Federation.

Conflicts of Interest: The authors declare no conflict of interest.

References

1. Davidson, D.W.; Handa, Y.P.; Ratcliffe, C.I.; Tse, J.S.; Powell, B.M. The ability of small molecules to form clathrate hydrates of structure II. *Nature* **1984**, *311*, 142–143. [CrossRef]

2. Kirchner, M.T.; Boese, R.; Billups, W.E.; Norman, L.R. Gas Hydrate Single-Crystal Structure Analyses. *J. Am. Chem. Soc.* **2004**, *126*, 9407–9412. [CrossRef] [PubMed]

3. Kirchner, M.T.; Bläser, D.; Boese, R. Co-crystals with Acetylene: Small Is not Simple! *Chem. Eur. J.* **2010**, *16*, 2131–2146. [CrossRef] [PubMed]

4. Antipin, M.Y. Low-temperature X-ray diffraction analysis: Possibilities in the solution of chemical problems. *Russ. Chem. Rev.* **1990**, *59*, 607–626. [CrossRef]

5. Chopra, D.; Row, T.G. In situ Cryocrystallization: Pathways to study intermolecular interactions. *J. Indian Inst. Sci.* **2007**, *87*, 167.

6. Otálora, F.; Gavira, J.A.; Ng, J.D.; García-Ruiz, J.M. Counterdiffusion methods applied to protein crystallization. *Prog. Biophys. Mol. Biol.* **2009**, *101*, 26–37. [CrossRef] [PubMed]

7. McPherson, A.; Cudney, B. Optimization of crystallization conditions for biological macromolecules. *Acta Crystallogr. Sect. F Struct. Biol. Commun.* **2014**, *70*, 1445–1467. [CrossRef]

8. Fedyanin, I.V.; Smol'yakov, A.F.; Lyssenko, K.A. In situ crystallization of 2,2-dimethoxypropane and dimethyldimethoxysilane: Hunting for Group 14 isomorphism. *Mendeleev Commun.* **2019**, *29*, 531–533. [CrossRef]

9. Jayatilaka, D.; Grimwood, D.J. Tonto: A Fortran Based Object-Oriented System for Quantum Chemistry and Crystallography. In *Computational Science—ICCS 2003*; Sloot, P.M.A., Abramson, D., Bogdanov, A.V., Gorbachev, Y.E., Dongarra, J.J., Zomaya, A.Y., Eds.; Springer: Berlin/Heidelberg, Germany, 2003; Volume 2660, pp. 142–151. ISBN 978-3-540-40197-1.

10. Bacon, G.E.; Curry, N.A.; Wilson, S.A.; Spence, R. A crystallographic study of solid benzene by neutron diffraction. *Proc. R. Soc. Lond. Ser. Math. Phys. Sci.* **1964**, *279*, 98–110.

11. Kahn, R.; Fourme, R.; André, D.; Renaud, M. Crystal structure of cyclohexane I and II. *Acta Crystallogr. B* **1973**, *29*, 131–138. [CrossRef]

12. Cox, E.G. The Crystalline Structure of Benzene. *Proc. R. Soc. Lond. Ser. Contain. Pap. Math. Phys. Character* **1932**, *135*, 491–498. [CrossRef]

13. Hassel, O.; Kringstad, H. *Tidsskr. Kjemi Og Bergves.* **1930**, *10*, 128–130.

14. Lonsdale, K.; Smith, H. LXII. Crystal structure of cyclohexane at −180 °C. *Lond. Edinb. Dublin Philos. Mag. J. Sci.* **1939**, *28*, 614–616. [CrossRef]

15. Price, W.C. The Structure of Diborane. *J. Chem. Phys.* **1947**, *15*, 614. [CrossRef]

16. Price, W.C. The Absorption Spectrum of Diborane. *J. Chem. Phys.* **1948**, *16*, 894–902. [CrossRef]

17. Lipscomb, W.N.; Streib, W.E. Growth, orientation, and X-ray diffraction of single crystals near liquid helium temperatures. *Proc. Natl. Acad. Sci. USA* **1962**, *48*, 911–913.
18. Smith, H.W.; Lipscomb, W.N. Single-Crystal X-Ray Diffraction Study of β-Diborane. *J. Chem. Phys.* **1965**, *43*, 1060–1064. [CrossRef]
19. Jordan, T.H.; Smith, H.W.; Streib, W.E.; Lipscomb, W.N. Single-Crystal X-Ray Diffraction Studies of α-N$_2$ and β-N$_2$. *J. Chem. Phys.* **1964**, *41*, 756–759. [CrossRef]
20. Jordan, T.H.; Streib, W.E.; Smith, H.W.; Lipscomb, W.N. Single-crystal studies of β-F$_2$ and of γ-O$_2$. *Acta Crystallogr.* **1964**, *17*, 777–778. [CrossRef]
21. Trotter, J. The crystal structure of nitrobenzene at −30 °C. *Acta Crystallogr.* **1959**, *12*, 884–888. [CrossRef]
22. Burbank, R.D.; Bensey, F.N. The Structures of the Interhalogen Compounds. I. Chlorine Trifluoride at −120 °C. *J. Chem. Phys.* **1953**, *21*, 602–608. [CrossRef]
23. Eyer, A.; Nitsche, R.; Zimmermann, H. A double-ellipsoid mirror furnace for zone crystallization experiments in spacelab. *J. Cryst. Growth* **1979**, *47*, 219–229. [CrossRef]
24. Brodalla, D.; Mootz, D.; Boese, R.; Osswald, W. Programmed crystal growth on a diffractometer with focused heat radiation. *J. Appl. Crystallogr.* **1985**, *18*, 316–319. [CrossRef]
25. Boese, R.; Antipin, M.Y.; Nussbaumer, M.; Bläser, D. The molecular and crystal structure of 4-methoxybenzylidene-4′-*n*-butylaniline (MBBA) at −163 °C. *Liq. Cryst.* **1992**, *12*, 431–440. [CrossRef]
26. Boese, R.; Bläser, D.; Nussbaumer, M.; Krygowski, T.M. Low temperature crystal and molecular structure of nitrobenzene. *Struct. Chem.* **1992**, *3*, 363–368. [CrossRef]
27. Boese, R.; Nussbaumer, M. In situ crystallisation techniques. *Oxf. Univ. Press* **1994**, *7*, 20.
28. Bruckmann, J.; Krüger, C. Chelating organophosphines: The use of comparative structural investigations to determine ligand properties. *J. Organomet. Chem.* **1997**, *536–537*, 465–472. [CrossRef]
29. Bruckmann, J.; Krüger, C. Tris(n-butyl)phosphine, Tris(tert-butyl)phosphine and Tris(trimethylsilyl)phosphine. *Acta Crystallogr. C* **1995**, *51*, 1152–1155. [CrossRef]
30. Waerder, B.; Pieper, M.; Körte, L.A.; Kinder, T.A.; Mix, A.; Neumann, B.; Stammler, H.-G.; Mitzel, N.W. A Neutral Silicon/Phosphorus Frustrated Lewis Pair. *Angew. Chem. Int. Ed.* **2015**, *54*, 13416–13419. [CrossRef]
31. Waerder, B.; Steinhauer, S.; Bader, J.; Neumann, B.; Stammler, H.-G.; Vishnevskiy, Y.V.; Hoge, B.; Mitzel, N.W. Pentafluoroethyl-substituted α-silanes: Model compounds for new insights. *Dalton Trans.* **2015**, *44*, 13347–13358. [CrossRef]
32. Steinfink, H.; Post, B.; Fankuchen, I. The crystal structure of octamethyl cyclotetrasiloxane. *Acta Crystallogr.* **1955**, *8*, 420–424. [CrossRef]
33. Patnode, W.; Wilcock, D.F. Methylpolysiloxanes. *J. Am. Chem. Soc.* **1946**, *68*, 358–363. [CrossRef]
34. Hoffman, J.D. Thermal and dielectric study of octamethylcyclotetrasiloxane. *J. Am. Chem. Soc.* **1953**, *75*, 6313–6314. [CrossRef]
35. Arzumanyan, A.V.; Goncharova, I.K.; Novikov, R.A.; Milenin, S.A.; Boldyrev, K.L.; Solyev, P.N.; Tkachev, Y.V.; Volodin, A.D.; Smol'yakov, A.F.; Korlyukov, A.A.; et al. Aerobic Co or Cu/NHPI-catalyzed oxidation of hydride siloxanes: Synthesis of siloxanols. *Green Chem.* **2018**, *20*, 1467–1471. [CrossRef]
36. Haas, M.; Christopoulos, V.; Radebner, J.; Holthausen, M.; Lainer, T.; Schuh, L.; Fitzek, H.; Kothleitner, G.; Torvisco, A.; Fischer, R.; et al. Branched Hydrosilane Oligomers as Ideal Precursors for Liquid-Based Silicon-Film Deposition. *Angew. Chem. Int. Ed.* **2017**, *56*, 14071–14074. [CrossRef] [PubMed]
37. Pelzer, S.; Neumann, B.; Stammler, H.-G.; Ignat'ev, N.; Hoge, B. Synthesis of Bis(pentafluoroethyl)germanes. *Chem.—Eur. J.* **2016**, *22*, 4758–4763. [CrossRef] [PubMed]
38. Pelzer, S.; Neumann, B.; Stammler, H.-G.; Ignat'ev, N.; Hoge, B. Synthesis of Tris(pentafluoroethyl)germanes. *Chem.—Eur. J.* **2016**, *22*, 3327–3332. [CrossRef]
39. Pelzer, S.; Neumann, B.; Stammler, H.-G.; Ignat'ev, N.; Hoge, B. The Bis(pentafluoroethyl)germylene Trimethylphosphane Adduct (C$_2$F$_5$)$_2$ Ge·PMe$_3$: Characterization, Ligand Properties, and Reactivity. *Angew. Chem. Int. Ed.* **2016**, *55*, 6088–6092. [CrossRef]
40. Arkhipov, D.E.; Lyubeshkin, A.V.; Volodin, A.D.; Korlyukov, A.A. Molecular Structures Polymorphism the Role of F . . . F Interactions in Crystal Packing of Fluorinated Tosylates. *Crystals* **2019**, *9*, 242. [CrossRef]
41. Solyntjes, S.; Neumann, B.; Stammler, H.-G.; Ignat'ev, N.; Hoge, B. Difluorotriorganylphosphoranes for the Synthesis of Fluorophosphonium and Bismuthonium Salts. *Eur. J. Inorg. Chem.* **2016**, *2016*, 3999–4010. [CrossRef]

42. Yokoyama, Y.; Ohashi, Y. Crystal and Molecular Structures of RCH$_2$CH$_2$SCH$_3$ (R = OCH$_3$, SCH$_3$). *Bull. Chem. Soc. Jpn.* **1998**, *71*, 1565–1571. [CrossRef]

43. Yokoyama, Y.; Ohashi, Y. Crystal and Molecular Structures of Methoxy and Methylthio Compounds. *Bull. Chem. Soc. Jpn.* **1999**, *72*, 2183–2191. [CrossRef]

44. Cutín, E.H.; Védova, C.O.D.; Mack, H.-G.; Oberhammer, H. Conformation and gas-phase structure of difluorosulphenylimine cyanide, F$_2$S(O)NCN. *J. Mol. Struct.* **1995**, *354*, 165–168. [CrossRef]

45. Boese, R.; Cutin, E.H.; Mews, R.; Robles, N.L.; Della Védova, C.O. ((Fluoroformyl)imido)sulfuryl Difluoride, FC(O)NS(O)F$_2$: Structural, Conformational, and Configurational Properties in the Gaseous and Condensed Phases. *Inorg. Chem.* **2005**, *44*, 9660–9666. [CrossRef] [PubMed]

46. Thalladi, V.R.; Weiss, H.-C.; Bläser, D.; Boese, R.; Nangia, A.; Desiraju, G.R. C−H···F Interactions in the Crystal Structures of Some Fluorobenzenes. *J. Am. Chem. Soc.* **1998**, *120*, 8702–8710. [CrossRef]

47. Desiraju, G.R.; Parthasarathy, R. The nature of halogen···halogen interactions: Are short halogen contacts due to specific attractive forces or due to close packing of nonspherical atoms? *J. Am. Chem. Soc.* **1989**, *111*, 8725–8726. [CrossRef]

48. Wheatley, P.J. The crystal and molecular structure of s-triazine. *Acta Crystallogr.* **1955**, *8*, 224–226. [CrossRef]

49. Coppens, P. Comparative X-Ray and Neutron Diffraction Study of Bonding Effects in s-Triazine. *Science* **1967**, *158*, 1577–1579. [CrossRef]

50. Bertinotti, F.; Giacomello, G.; Liquori, A.M. The structure of heterocyclic compounds containing nitrogen. I. Crystal and molecular structure of s-tetrazine. *Acta Crystallogr.* **1956**, *9*, 510–514. [CrossRef]

51. Wheeler, G.L.; Colson, S.D. Intermolecular interactions in polymorphic *p*-dichlorobenzene crystals: The α, β, and γ phases at 100 °K. *J. Chem. Phys.* **1976**, *65*, 1227–1235. [CrossRef]

52. Wheeler, G.L.; Colson, S.D. γ-Phase ıt p-dichlorobenzene at 100 K. *Acta Crystallogr. Sect. B* **1975**, *31*, 911–913. [CrossRef]

53. Boese, R.; Kirchner, M.T.; Dunitz, J.D.; Filippini, G.; Gavezzotti, A. Solid-State Behaviour of the Dichlorobenzenes: Actual, Semi-Virtual and Virtual Crystallography. *Helv. Chim. Acta* **2001**, *84*, 1561–1577. [CrossRef]

54. Gavezzotti, A.; Filippini, G. Geometry of the Intermolecular X-H···Y (X, Y = N, O) Hydrogen Bond and the Calibration of Empirical Hydrogen-Bond Potentials. *J. Phys. Chem.* **1994**, *98*, 4831–4837. [CrossRef]

55. Dikundwar, A.G.; Sathishkumar, R.; Guru Row, T.N.; Desiraju, G.R. Structural Variability in the Monofluoroethynylbenzenes Mediated through Interactions Involving "Organic" Fluorine. *Cryst. Growth Des.* **2011**, *11*, 3954–3963. [CrossRef]

56. Murray, J.S.; Paulsen, K.; Politzer, P. Molecular surface electrostatic potentials in the analysis of non-hydrogen-bonding noncovalent interactions. *Proc. Indian Acad. Sci. Chem. Sci.* **1994**, *106*, 267–275.

57. Dikundwar, A.G.; Guru Row, T.N. Evidence for the "Amphoteric" Nature of Fluorine in Halogen Bonds: An Instance of Cl···F Contact. *Cryst. Growth Des.* **2012**, *12*, 1713–1716. [CrossRef]

58. Dikundwar, A.G.; Guru Row, T.N. Tracing a Crystallization Pathway of an RT Liquid, 4-Fluorobenzoyl Chloride: Metastable Polytypic Form as an Intermediate Phase. *Cryst. Growth Des.* **2014**, *14*, 4230–4235. [CrossRef]

59. Wang, H.; Xiao, H.; Liu, N.; Zhang, B.; Shi, Q. Three New Compounds Derived from Nitrofurantoin: X-Ray Structures and Hirshfeld Surface Analyses. *Open J. Inorg. Chem.* **2015**, *5*, 63–73. [CrossRef]

60. Dikundwar, A.G.; Sathishkumar, R.; Guru, R.T.N. Fluorine prefers hydrogen bonds over halogen bonds! Insights from crystal structures of some halofluorobenzenes. *Z. Für Krist. Cryst. Mater.* **2014**, *229*, 609–624. [CrossRef]

61. Masters, S.L.; Mackie, I.D.; Wann, D.A.; Robertson, H.E.; Rankin, D.W.H.; Parsons, S. Unusual asymmetry in halobenzenes, a solid-state, gas-phase and theoretical investigation. *Struct. Chem.* **2011**, *22*, 279–285. [CrossRef]

62. Nath, N.K.; Naumov, P. In situ crystallization and crystal structure determination of chlorobenzene. *Maced. J. Chem. Chem. Eng.* **2015**, *34*, 63–66. [CrossRef]

63. Vasylyeva, V.; Catalano, L.; Nervi, C.; Gobetto, R.; Metrangolo, P.; Resnati, G. Characteristic redshift and intensity enhancement as far-IR fingerprints of the halogen bond involving aromatic donors. *CrystEngComm* **2016**, *18*, 2247–2250. [CrossRef]

64. Klapötke, T.M.; Krumm, B.; Moll, R.; Rest, S.F.; Vishnevskiy, Y.V.; Reuter, C.; Stammler, H.-G.; Mitzel, N.W. Halogenotrinitromethanes: A Combined Study in the Crystalline and Gaseous Phase and Using Quantum Chemical Methods. *Chem. Eur. J.* **2014**, *20*, 12962–12973. [CrossRef] [PubMed]

65. Golovina, N.I.; Atovmyan, L.O. Crystal structure of iodotrinitromethane. *J. Struct. Chem.* **1967**, *7*, 230–233. [CrossRef]

66. Göbel, M.; Tchitchanov, B.H.; Murray, J.S.; Politzer, P.; Klapötke, T.M. Chlorotrinitromethane and its exceptionally short carbon–chlorine bond. *Nat. Chem.* **2009**, *1*, 229–235. [CrossRef]

67. Berrueta Martínez, Y.; Reuter, C.G.; Vishnevskiy, Y.V.; Bava, Y.B.; Picone, A.L.; Romano, R.M.; Stammler, H.-G.; Neumann, B.; Mitzel, N.W.; Della Védova, C.O. Structural Analysis of Perfluoropropanoyl Fluoride in the Gas, Liquid, and Solid Phases. *J. Phys. Chem. A* **2016**, *120*, 2420–2430. [CrossRef]

68. Jelsch, C.; Soudani, S.; Ben Nasr, C. Likelihood of atom–atom contacts in crystal structures of halogenated organic compounds. *IUCrJ* **2015**, *2*, 327–340. [CrossRef]

69. Hunter, C.A. Arene—Arene Interactions: Electrostatic or Charge Transfer? *Angew. Chem. Int. Ed. Engl.* **1993**, *32*, 1584–1586. [CrossRef]

70. Salonen, L.M.; Ellermann, M.; Diederich, F. Aromatic Rings in Chemical and Biological Recognition: Energetics and Structures. *Angew. Chem. Int. Ed.* **2011**, *50*, 4808–4842. [CrossRef]

 *crystals*

Review

Intermolecular Interactions in Functional Crystalline Materials: From Data to Knowledge

Anna V. Vologzhanina

A. N. Nesmeyanov Institute of Organoelement Compounds, RAS. 28 Vavilova street, 119991 Moscow, Russia; vologzhanina@mail.ru

Received: 26 August 2019; Accepted: 12 September 2019; Published: 13 September 2019

Abstract: Intermolecular interactions of organic, inorganic, and organometallic compounds are the key to many composition–structure and structure–property networks. In this review, some of these relations and the tools developed by the Cambridge Crystallographic Data Center (CCDC) to analyze them and design solid forms with desired properties are described. The potential of studies supported by the Cambridge Structural Database (CSD)-Materials tools for investigation of dynamic processes in crystals, for analysis of biologically active, high energy, optical, (electro)conductive, and other functional crystalline materials, and for the prediction of novel solid forms (polymorphs, co-crystals, solvates) are discussed. Besides, some unusual applications, the potential for further development and limitations of the CCDC software are reported.

Keywords: cambridge structural database; crystal structures; knowledge-based analysis; intermolecular interactions; structure–property relations; supramolecular chemistry

1. Introduction

The history of investigations devoted to the analysis of networks between chemical composition, molecular, and crystal structures and numerous properties of compounds dates back to 1960s. The development of the X-ray diffraction technique and computational routines allowed to collect information about crystal structures of plenty of inorganic, organic, organometallic, and macromolecular compounds. Early findings in the field of composition–structure–properties networks of these solids gave us knowledge about the typical molecular geometry [1,2], steric, and electronic effects of functional groups [3,4], principles of molecular packing [5], role and energetic of numerous intermolecular interactions [6,7]. Longstanding efforts of the crystallographic community to present crystallographic data in a machine-processed format, and to collect these data in crystallographic databases combined with recent progress in software development promotes further insights into the synthesis of novel solids with desired physicochemical properties (including optical, magnetic, electrical, mechanical, and others).

The literature contains numerous review articles devoted to recent advances in crystal engineering [8,9], design of functional organic (see, for example, Refs. [10–13]) and inorganic [14,15] materials, and the development of software for analysis of molecular crystals [16–18]. However, in my opinion knowledge-based analysis and corresponding software are still insufficiently used by chemists, who often analyze relations between functional properties and intermolecular interactions on the level of bond geometry and Figures of crystal packing. On the other hand, analysis of applications of the software by the end users of software can help software developers to find and overcome limitations of their algorithms, and to propose lines for further development. Thus, in the present paper, to demonstrate the effectiveness of knowledge-based analysis of structure–property relations in crystals, some of the relations, possible application of data–knowledge studies, and predictions to analyze them, and a brief description of procedures will be described. Among a huge number of papers published in these fields, the manuscript cites only (*i*) recent works devoted to the analysis of correlations between intermolecular interactions and properties of a small molecule (for example its inclination to form polymorphs, solvates, and co-crystals), or a corresponding solid (from a well-known requirement for non-linear optical materials to crystallize in acentric space groups, to recent studies devoted to the effect of solvent presence on mechanical properties), and (*ii*) the corresponding software developed to investigate these correlations and to design novel solid forms with desired physicochemical properties. As the description of structure–property networks describes mainly the papers published in the last 10 years in the field of organic, organometallic, and coordination crystals, then the software under discussion will be limited with those developed by the Cambridge Crystallographic Data Centre (CCDC) for material chemistry and crystallography. Various examples of applications of the CCDC software to functional materials including their combination with other software, and restrictions found, will be given.

First, the Cambridge Structural Database (CSD) and components of the CSD-Enterprise will be described in Section 2. Then, some properties related to the appearance of a given supramolecular associate, and the tools to search for an associate in the CSD will be reported in Section 3. The properties of solids dependent on the crystal morphology, the Bravais, Friedel, Donnay, and Harker (BFDH) tool for crystal morphology prediction and its' application to affect crystal morphology, polymorphism, and solvatomorphism will be reported in Section 4. Knowledge-based predictions of H-bonded polymorphism, co-crystal formation, mapping of likely intermolecular interactions, and conformer generation for the synthesis of novel functional materials will be discussed in Section 5, and the analysis of local connectivity and whole architectures of solvent molecules in Section 6. Finally, Section 7 contains information about Python API algorithms compatible with the CSD-Enterprise and about some examples of the successful combination of the CSD-Enterprise tools with external software applicable for investigations in the field of structure–property networks.

2. The Cambridge Structural Database (CSD) and its' Libraries and Modules

The Cambridge Structural Database [19,20] is one of several databases containing crystal structures of various compounds whose structures have been determined using crystallographic techniques. Particularly, the CSD contains crystal structures of organic, and organoelement compounds, and metal–organic complexes. These data are collected from publications all other the world combined with structural determinations with no accompanying manuscript. Single-crystal data are included to the CSD even if no coordinates are available, while powder studies are included only if cell parameters, atomic coordinates, and refinement details were reported. The original value of the CSD was to simplify access to individual structures, to help crystallographers to avoid redetermination of previously reported structures, and to allow easy sharing of work within the chemical and crystallographic community. In 2019, the number of the CSD individual entries overpassed 1,000,000! This value can be compared with ca. 200,000 of inorganic crystal structures from the Inorganic Crystal Structures Database (ICSD) [21], or 150,000 entries from the Protein Data Bank (PDB) [22].

The second benefit of the database comes from data–knowledge analysis derived from the entire collection. Since the late 1960s, analysis of crystallographic data allowed us to demonstrate a non-uniform distribution of crystal systems and space groups, to estimate ionic and van-der-Waals radii of various elements, or to propose close packing of molecules within a crystal. Analysis of intermolecular distances to reveal structure-forming interactions has been known for tens of years; the role of hydrogen bonds, π...π stacking, and halogen...halogen interactions can not be overestimated [6,7]. Nowadays, information about the crystal structures of discrete organic and organometallic compounds allows us to reformulate principles of their packing [23,24], and to investigate intermolecular interactions more thoroughly. Not only information about geometry and frequency of occurrence of an unusual interaction, synthon, or tecton is extracted from the CSD to support a conclusion about its' role in crystal packing and electronic effects that cause its' appearance [25–29]; but it is also recommended to additionally compare interaction occurrence relative to what would be expected at random [30–33].

Data about molecular geometry—from bond lengths and angles to torsions and ring geometry—form the basis of our understanding of the energetics of molecular conformation. Analysis of these data allows us to understand not only the most stable conformations, but also to rationalize steric and electronic effects of substitutents, or co-formers on the geometry, or to link it with a reaction pathway [34–36].

Taking the value of the conformational preferences exhibited by molecular fragments, and intermolecular interactions into account, two dynamic knowledge-based libraries were derived from the CSD. These are: *Mogul* [37], which stores intramolecular information, and *IsoStar* [38], which collects information about intermolecular interactions. *Mogul* contains information about millions of bond lengths, angles, and torsions, each relating to a specific chemical environment. Data from this library can be organized into distributions providing a click-of-a-button access to structures used to construct the distribution. Besides, one can check the geometry of a novel structure (this is of importance for structure refinement from powder X-Ray diffraction data, or for understanding the environmental effects of combinative factors and steric exclusion) or to check the geometry of a protein–ligand docking pose. Data for *Isostar* were collected not only from the CSD, but also from protein–ligand complexes in the PDB having a resolution better than 2 Å, and ab initio data of the key interactions. *Isostar* provides the searchable distribution of contact functional groups around a central one (such distributions can be obtained for either the CSD or the PDB collections), and data about geometry (and sometimes energy) of various interactions.

Thus, the CSD and associated libraries serve also as sources of supramolecular knowledge for applications software that addresses specific problems in structural chemistry, rational drug design and crystal engineering. In accord with these tasks, various modules comprising the CSD-Enterprise suite are under constant development (Figure 1). The CSD-System contains the CSD, *Mogul*, and *Isostar* databases; *ConQuest* and *webCSD* provide off-line and on-line searches within the CSD; *Mercury* and *Hermes* are the CCDC's 3D vizualizers of, respectively, small and macromolecular crystal structures. Currently, both *Hermes* and *Mercury* are linked with some CSD-Materials and the CSD-Discovery tools of the CSD-Enterprise. The CSD-Materials module serves for use in crystal engineering and materials design, as well as for the structure solution from powder XRD data with *DASH package*. Various tools and programs for drag analysis and discovery are combined within the CSD-Discovery module. Note, that the *Full Interaction Map* (FIM) tool and *Conformer Generator* tool are intensively used for both material chemistry and drug design. At last, every scientist from this community is welcome to install previously published or to write (and distribute) his/her own CSD Python API scripts.

Figure 1. Modules comprising the Cambridge Structural Database (CSD)-Enterprise (ver. 2019).

The CSD-Materials Module

The main CSD module applicable for investigation and comparison of solid forms and intermolecular interactions, which govern packing of these solids, is the CSD-Materials module. Its' components in the CSD-2019 version allow us to perform:

- Analysis of H-bond motifs (searching of motifs and statistics of their occurrences; assessing the risk of polymorphism via H-bond propensities; prediction of co-crystal formation).
- Analysis of crystal packing features (searching on selected motifs, analysis of packing similarity and building of a packing similarity tree diagram).
- Calculation of theoretical crystal morphology, gas phase MOPAC (Molecular Orbital PACkage) calculation and 'UNI' (UNIversal) force field intermolecular energy calculations.
- Mapping of interaction preferences around an isolated molecule.
- Analysis of solvate and hydrate crystal structures (searching and classification of H-bond motifs, calculation of the volume occupied in a unit cell with solvent molecules, mapping of interaction preferences around them).
- Generation of conformers based on geometrical statistics from the CSD.
- Solution of crystal structures from powder diffraction data using *DASH*.

Although this software was developed and attested mainly for pharmaceutical crystal forms, numerous examples of structure–property networks found for other organic and organometallic compounds with valuable optical, electrochemical, mechanical, and other properties are given below. These studies demonstrate that the possible range of applications of the CSD-materials module can be much expanded, although its application to coordination polymers is less diverse.

3. Search and Analysis of Supramolecular Associates

3.1. Search on Hydrogen-Bond Motifs

The utility of big-data analysis of intermolecular interactions extracted from crystallographic databases is based on the fact that the frequency distributions of functional group contacts in the CSD are directly related to the corresponding interaction energies in solution [39]. Analysis of crystal packing in pharmaceuticals, their homologues, polymorphs, co-crystals, and solvates gives one information about preferable synthons that could appear in a binding pocket occupied with this drug (although some weak interactions can be over-, or under-estimated). Data knowledge about typical water associates in crystals of small molecules is important for biochemistry as similar hydrate architectures were found in crystals of inorganic, organic, and macromolecular compounds [40,41].

Among all contacts, H-bonds are the strongest and the most directional interactions that play the dominant role in the crystallization and stability of organic solids, thus their analysis is a central theme of crystal engineering [42–44]. For example, stable H-bonding motifs are used widely to fix olefins in photoreactive positions for [2+2] cycloaddition [45,46]. Analysis of H-bonding is important for understanding proton conductivity [47] and high-temperature ferroelectricity caused by proton tautomerism in polar space groups [48–51]. Reversal of an electric field in the latter solids causes a switch of O-H...O or N-H...N bonds to O...H-O or N...H-N ones accompanied with ketone/enol [48] or neutral/zwitterion [49] isomerisation or imidazole tautomerization [50,51] and the polarity reversal.

Comparison of various H-bonded motifs found between similar functional groups allows one not only to estimate which of them is more stable and abundant to occur but also to shed light on stereoelectronic effects stabilizing various associates [29,52,53]. For example, analysis of 23 phenylglycine amide benzaldimines revealed only five types of H-bonded motifs; the choice of the particular motif depends on the number of H-bonding donors and acceptors, the ease with which the motif is formed, and the possibility of the motif to accommodate additional substituents [54]. The stability of polymorphs of a multifunctional molecule is known to be determined by the energy of an H-bonded motif [55]—that is of importance for drug and food industry, production of high-energy compounds, and dyes and pigments. For these materials the presence of uncommon weak H-bonds in an observed solid form can be indicative of the potential for alternative packing to form another polymorph with stronger interactions [56]. Instead, for high-energy compounds strong H-bonded interactions are disadvantageous as these can prevent molecules from dense packing [57–59]. The appearance of particular H-bonded motifs (spiral chains, bilayers, and others) is characteristic for spontaneous resolution of racemic and nonracemic mixtures of some chiral compounds [60–62] at crystallizations. Other applications of the analysis of H-bonded associates include salt formation via evaluating synthon competition [63,64], assessing putative structures from crystal structure prediction [65], and analysis of H-bonding in drugs and vitamins [66].

Thus, the search for all representatives of a given H-bonded motif, or for all H-bonded motifs found between some functional groups within the CSD, as well as their classification and statistics of occurrence, become of interest. All these possibilities are realized within the *CSD Motif Search* (Figure 2). A user can select a pre-defined motif from a special library or generate a new motif. To generate a motif, one should sketch a functional group, select the atoms of the functional group which define the contact(s) of the motif, define interatomic distance, and select the type of motif (a ring, an infinite, or a discrete chain) and the number of contacts within the motif. The search can be carried out in the current version of the CSD, in individual refcodes and families of refcodes or in files of structures. The results can be viewed by motifs with the number of hits found and the frequency of occurrence or by structures with the motifs found for each hit. As more than one motif can be found in a crystal structure, their combination can give not only ring and chain motifs but also H-bonded layers and frameworks. Unfortunately, classification of all matches in accord with all non-equivalent H-bonding associates is unavailable in the current version. Besides, the notation used to describe these motifs [67] do not correspond to that recommended to describe underlying nets of crystalline networks and clusters [68–71]. Analysis of H-bonded motifs is part of the *H-bond propensity* tool, the *Co-Crystal Formation* tool, and the *Hydrate Analyzer* described in Sections 5.1, 5.2 and 6, respectively.

Figure 2. Flowchart for the *CSD Motif Search*.

3.2. StudyingCrystal Packing Features

Besides robust and directed intermolecular interactions, such packing features as π...π stacking, dipole...dipole or halogen, pnicogen, and carbon bonding as well as molecular size-shape regularities can become investigation objects. Some properties of solids are associated with molecular packing instead of intermolecular interactions. These are such properties as luminescent properties of co-crystals of anticrowns with aromatic molecules and hydrocarbons [72–74], electroconductivity of layered donor-acceptor complexes [75,76], optical properties of π-conjugated molecules [77,78], dense packing of high-energetic compounds associated with detonation, and stability properties [59,79]. In this context, it becomes of interest how polytopic molecules or molecules with a lack of limited H-bonding functionalities pack, and which factors govern their packing. A comparison of polymorphs, solvates, and co-crystals of such compounds gives a key for better understanding of these factors. For example, an analysis of 88 crystal structures containing trimeric perfluoro-ortho-phenylen mercury (TPPM) revealed only four motifs of their packing with Lewis bases, governed mainly by Hg...C and π...π interactions [80]. It was shown that the type of supramolecular motif strongly correlates with the nature of a co-former. Another example of the role of weak interactions is the co-crystallization of diphenyl dichalcogenides Ph_2E_2 (E = Se, Te) with diiodotetrafluorobenzene that is incorporated during crystallization between stable tectonic chain Ph_2E_2 architectures, which were also found in the crystals of pure Ph_2E_2 [81]. The co-crystallization of this molecule can be considered as a replacement of E–E and Te–π(Ph) chalcogen bonds with I–E and I–π(Ph) halogen bonds via insertion of stacks of halogen bond donors between tectons of halogen acceptors. Analysis of intermolecular interactions by means of the energy framework diagrams revealed that the energies of corresponding chalcogen and halogen bonds were lower than those between the stacks of pure Ph_2E_2 and diiodotetrafluorobenzene, and illustrates an approach to binary crystals formed by matching tectons (that can be found with the *Crystal Packing Features* tool).

Analysis of mutual disposition of molecules, in this case, can be carried out with the *Crystal Packing Features* tool of the CSD-Materials module. In contrast with the *CSD Motif Search*, a "packing feature" can be generated only from a displayed structure by selecting the atoms and bonds to define the feature. A search query is constructed without sketching any functional groups and atoms based on selected atoms, bonds, and intermolecular distances. Then the CSD is searched to identify crystal structures that contain a similar mutual disposition of atoms. The hits are automatically overlaid with the original geometry, and root-mean-square distance (RMSD) value is reported as a measure of packing similarity. For example, the binary stacks of TPPM and aromatic molecules can be found in a series of co-crystals with luminescent properties [72]. This motif, one of four typical for TPPM co-crystals [80], can be constructed from a TPPM molecule and two aromatic C_6 rings situated above and below the TPPM meanplane (Figure 3a). The *Crystal Packing Feature* search allows us to extract from the CSD other examples of such potentially luminescent co-crystals, and to compare the distances between the meanplanes of planar molecules as a measure of charge transfer from electron-rich aromatic molecules to anticrowns.

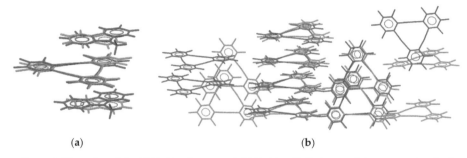

(a)　　　　　　　　　　　　　　　(b)

Figure 3. Examples of the (**a**) *Packing Feature* and (**b**) *Packing Similarity* search hits. A sandwich of a TPPM with two aromatic C_6 rings was constructed as the *Packing Feature* from a reference (red) {MOXMIV}[a] molecule and was also found in (blue) {QATTAH}. The *Packing Similarity* comparison of an orthorhombic polymorph of TPPM (blue) {MOXMAN02} with its' three monoclinic polymorphs gives the best similarity with (red) {MOXMAN03}. [a] Here and below a six-letter CSD-Refcode of a compound is given in Figure braces.

A strategy to co-crystal formation based on packing similarity between various co-formers was suggested in Ref. [82]. In this paper, the authors suggested replacing the anion in the crystal of lamaviudine saccharinate with another anion having a similar disposition of acceptor groups (the nature of atoms within acceptors was allowed to vary). Crystal structure and intermolecular interactions of lamaviudine maleate taken as a hit of the CSD search, indeed, were very similar to that of saccharinate. This tool can also be applied to intramolecular interactions or mutial disposition of more than two molecules. For example, the analysis of semicarbazone conformations distinguished that the syn–anti–syn–syn conformation predominates the syn–anti–anti–anti one [83].

3.3. Crystal Packing Similarity Tool

While the *Crystal Packing Features* described above allows one to select from the CSD all supramolecular associates with a user defined mutual disposition of molecules, the *Crystal Packing Similarity* tool can be used to compare a large set of solids containing one co-former and to measure similarity between them, as well as to obtain similarly overlaid molecular packing automatically. This tool was used for comparison of the experimental crystal structures with those predicted during the sixth blind test of organic crystal structure prediction methods [84]. Such investigations are also of great importance for the analysis of isostructural compounds [85] and some properties that occur only within a given group of supramolecular associates. For example, tetraphenylborates of *N*-salicylideneanilines form isostructural series of photochromic solids due to the packing of cations within a cavity formed by phenyl groups of anions, while more dense packing in other salts gives non-photochromic solids [86]. The effect of substituents on isostructurality of compounds was widely analyzed for both rigid [87–92] and flexible molecules [93–98]. Not only long-accepted isostructurality of Cl- and Me- or -Cl, -Br, and -I substituted molecules were revealed, but also the equivalence of ethylene and azo-bridges [92], or that of azide and iodide substituents [85] were demonstrated. Besides the *Crystal Packing Similarity* tool, XPac [99] and Crycom [100] are also among the software appropriate to carry out such a comparison.

The *Crystal Packing Similarity tool* is applicable to compare multiple structures containing the same compound, e.g., its polymorphs, hydrates, solvates, co-crystals, and salts. Within this method molecular clusters (typically containing a central molecule and its' 14 closest neighbors) are built for each structure, and the clusters are compared with some geometric tolerance to define whether packing is similar or not. Small differences in packings (variation of halogen atoms, absence of some hydrogen atoms, presence of few independent molecules and small co-formers) can be ignored. The results of the comparison are grouped in accord with the number of neighbors forming molecular clusters (from 1 to 2 for different structures to all 14 for pseudo-isostructural compounds), and the groups can also be compared with each other to reveal synthons and associates common for different groups. Such grouping reflects the fact that similar motifs found in polymorphs and solvates typically keep similar intermolecular connectivity and energy of pair interactions [80,101]. The result of such analysis is the dendrogram visualizing crystal packing similarity between different groups of crystal structures. For example, analysis of 50 crystal structures, containing carbamazepin revealed three main motifs ("translation stacks", "inversion cup", and "co-former pairing", Figure 4), and none of these appears via H-bonding [102]. Instead, these motifs seem to represent the most efficient methods for packing of carbamazepin molecules while leaving the carboxamide group free to form hydrogen bonds. Note, that these results coincide with those obtained for carbamazepin-containing solids using the XPac method [103].One of the carbamazepin pseudo-isostructural groups is represented by metastable polymorph II and its solvates and hydrates situated inside the channels formed by the hydrophobic aromatic surfaces of this molecule, which were found to act as stabilizers of this solid [104]. These solvent molecules within the channels can even be replaced with hydrophobic polymer guest molecules [105].

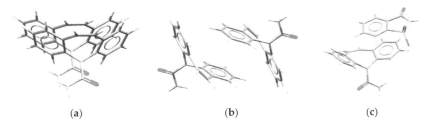

(a) (b) (c)

Figure 4. Carbamazepin molecules packed in (a) "translation stacks" in {CBMZPN11}, (b) "inversion cups" in {CBMZPN01} and (c) "co-former pairs" in {UNEZAO}.

Comparison of an orthorhombic polymorph of trimeric perfluoro-ortho-phenylen mercury with its' three monoclinic polymorphs gives up to six similarly packed molecules. Figure 3b visualizes that in crystals of {MOXMAN02} and {MOXMAN03} perfluoro-ortho-phenylen mercury form similar stacks, but these stacks are packed in different ways. The interplay of packing motifs and hydrogen bonds were demonstrated on the example of 37 enantiopure and racemic salts of methylethedrine, where the molecule forms six "isostructural" groups, two of which were characterized by alternating motifs of H-bonding [106]. Analysis of co-crystals of meloxicam with carboxylic acids reveal the acids, which are able to break H-bonded meloxicam dimers, and, therefore, to compare the effect of crystal packing on dissolution behavior [107]. The comparison of crystal structures of hydrates of furosemide: Nicotinamide co-crystal suggests the mechanism of dehydration [108]. Galcera, et al. demonstrated by using this tool that the isostructurality in co-crystals of lamotrigin with flexible dicarboxilic acids appears at the presence of solvents able to mimic the difference in conformation and volume of co-formers [109]. The analysis of 42 tyrammonium salts demonstrated that isostructural cation packing can occur even with structurally different anions, with different hydration states and with different hydrogen bonding; realization of different packing was associated with various conformations of the ethylammonium group of the cation [110]. Martins, et al. demonstrated that similarity between supramolecular clusters occurs only for molecules able to exhibit similarity in molecular shape-size parameters (thus called isostructural molecules), contact area, and energy of intermolecular interactions [111]. Fenamate co-crystals with 4,4'-bipyridine keep the packing arrangement of the initial solid connected by the 4,4'-bipyridine [112]. The *Crystal Packing Similarity* tool was also used to investigate 16 trospium chloride containing structures. This molecule realizes similar conformation in the majority of solvates including a sesquihydrate, while its conformation in dihydrate significantly varies from that in other solvates. All solvates but water belong to the same group as two trospium chloride polymorphs [113]. Dehydration and desolvation of lenalidomide follow "isostructurality" in crystal packing: All solvates at heating convert to its' thermodynamically stable anhydrous form, whereas all hydrates upon dehydration convert to a metastable anhydrous polymorph able to transform upon further heating to the stable polymorph [114]. Analysis of H-bonding and crystal packing similarity for the crystal structures of vitamin D analogs showed that various conformations of its A-ring are predetermined by H-bonding of hydroxyl group both in crystals and in the binding pocket of the vitamin D receptor and that the exocyclic methylene group also influences H-bonding pattern in solids [66].

To sum up "isostructurality" found by means of the *CSD Crystal Packing Similarity* tool ignores the requirement for identical symmetry of compounds, and unit cell parameters. Instead, the families of structures with matching molecular clusters when superimposed can be found to reveal crystal packing relationships, motif stability, and hydrogen-bond competition for a variety of solid forms of a molecule (and its' homologues or structural analogs).

3.4. CSD-Crossminer

The CSD-Crossminer is the novel software of the Cambridge Crystallographic Data Centre first appeared as part of the CSD-Enterprise in 2018. Its utility is based on ideas previously demonstrated on the example of carboxylate/tetrazolate [115], hydrate/peroxosovate [116], and maleate/saccharinate [82] behavior in solids and potential of 3D macrocyclic analogues of small-molecules to serve as drugs [117]. Particularly, molecules which form similar intermolecular interactions and have comparable size and shape tend to form similar supramolecular associates, and hence are able to form isotypical solvates, or bind with similar functional groups within a binding pocket of a macromolecule. A search defined within this software does not contain any chemical formulas or functional groups but uses such chemical feature as "H-bond donor", "H-bond acceptor", "hydrophobe", and others. Starting from a real molecule, mutual disposition of these chemical features can be defined, as well as the disposition of complementary functional features, which may come from other molecules, co-formers, solvates, or proteins and tolerance. The search is carried out within the CSD and the PDB databases and gives a number of hits potentially able to replace the reference molecule in solids or binding pockets. This software was developed as a pharmacophore query tool able to identify common protein binding sites in macromolecules, to determine structural motifs that are able to interact with similar binding sites, to estimate which ligand modifications are tolerated in a binding pocket, and others. More details about this tool are given in a white paper at *https://www.ccdc.cam.ac.uk/whitepapers/csd-crossminer-versatile-pharmacophore-query-tool-successful-modern-drug-discovery/*. However, the great potential of this tool for material design should be mentioned. Using this tool, molecular templates for constructions of polynuclear complexes or porous compounds, like zeolites, can be found in the CSD (this means that these molecules not only possess a desired conformation, but also were once synthesized and isolated). As pores of metal–organic frameworks are somehow equivalent with the binding pockets, the potential of metal-organic frameworks in respect to catalysis, separation of complex mixtures, or host of guest molecules, can probably be evaluated using this software.

4. Solid Form Calculations

Calculations included within the CSD-Materials module of Mercury allows one to simulate crystal morphology using the BFDH (Bravais, Friedel, Donnay, and Harker) method, to perform semi-empirical MOPAC calculations and to evaluate force-field intermolecular energy calculations using the 'UNI' intermolecular potentials. The MOPAC software is a semi-empirical quantum chemistry program based on NDDO (neglect of diatomic differential overlap) approximation, which allows a user to perform some calculations (geometry optimization, bond order calculations, molecular electrostatic potential visualization) for isolated molecules. The 2007 or 2009 version can be obtained free for academic users at *http://openmopac.net/*. The application of intermolecular potentials for analysis of interactions between two molecules, of a molecular cluster, or of the total packing energy is beyond the scope of this review and will not be described here. See the description of the empirical "UNI" potentials and their possible applications with Refs. [118,119]. The BFDH morphology tool instead will be described in detail below.

4.1. BFDH Morphology Prediction

Some of the macroscopic properties of solids depend on supramolecular architectures of symmetry elements formed by molecules and ions. Thus, if a supramolecular architecture is anisotropic, then, properties measured become anisotropic too; and their understanding and prediction require knowledge of disposition of functional groups, supramolecular synthons, or symmetry elements as compared with crystal faces. Having experimental data about Miller indices of crystal faces, one can examine corresponding planes using the free *Packing/Slicing tool* of Mercury. Particularly, indexing of the crystal phases becomes of practical interest for nonlinear optics [120], organic photonics [121], piezoelectrics [122–124], and organic electronics [125–128]. Knowledge of functional groups or supramolecular synthons forming crystal faces allowed in some cases to rationalize mechanical effects in dynamic crystals [129–132] such as self-healing, jumping, bending, twisting at heat, humidity, force or light, and to describe pressure-induced phase transitions (see Refs. [133–135] for analysis of such transitions in amino acids). Particle size and shape also affect material properties, such as tabletability [136], thus, their control is of special interest for the pharmaceutical industry.

Taking into account that the solvent can affect crystal morphology [137,138], it becomes of interest to rationalize solvent selection and to predict its' effect on crystal size and shape. Analysis of functional groups forming each surface gives a clue to solvent choice to prevent crystal growth in some directions, and vice versa, to rationalize the dissolution of a single crystal over various crystal faces [139]. For example, long-chain alkylboron capped tris-pyrazoloximates and clathrochelates readily form thin plate crystals with their main faces formed by hydrophobic alkyl groups (Figure 5) [140,141]. Their X-ray quality single crystals are obtained from polar solvents able to bind with small crystal faces formed by polar groups via hydrogen or halogen interactions, while crystallization from hydrocarbons typically results in twinned and turbostratic conglomerates of crystals. Similarly, ibuprofen [142] single crystals are faceted with (1 0 0), (0 0 2), and (0 1 1) faces, formed by hydrophobic, van der Waals and H-bonded interactions, respectively. As follows from the strength of intermolecular interactions in a crystal, and the strength of interactions between a solute and a solid [143], a polar protic solvent should bind to (0 0 2) faces to prevent the formation of needle morphology. This strategy based on the type of functional groups forming crystal faces was successfully used to optimize crystal morphology of lovastatin [144], 1-hydroxypyrene [145], isoniazid [146], tolbutamide [147] and N-benzyl-2-methyl-4-nitroaniline [148].

For phenacetin single crystals, the crash-cooling experiments resulted in the crystallization of needle-like crystals, while slow growth yielded similar single crystals with a hexagonal-like BFDH-predicted morphology affected by the solvent used [149]. Application of additive molecules able to bind with selected crystal faces also allows the of the controlling crystal morphology [150–153]. Yunqi Liu and coworkers suggested an interesting modification of such technique to affect the morphology and morphology-dependent breathing effect of metal–organic frameworks via the addition of a small amount of slightly modified ligands to the reaction mixture [154,155]. The hydrolysis of selected ligand decreased the growth of an anisotropic coordination polymer in some directions; the size of a crystal face containing pores was found to affect gas adsorption.

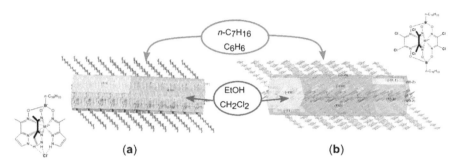

Figure 5. Bravais, Friedel, Donnay, and Harker (BFDH) predicted the morphology (blue) and hydrophobic groups forming the main crystal faces of a long-chain alkyl (a) tris-pyrazoloximate {1497845}, and (b) hexahalogenoclathrochelate {YUFYUV04}.

Having access to CSD-Materials, one can predict the morphology using the Bravais, Friedel, Donnay, and Harker (*BFDH*) crystal morphology tool (Figure 5). This method implies the idea that the crystals are preferable to grow along the direction with strong intermolecular interactions. Not only simulated but also experimental morphology can be depicted and analyzed to evaluate functional groups which form a crystal face and to estimate preferable interactions which occur between this face, a solvent, or an additive. Crystal morphology prediction can be used to estimate the direction of a preferred orientation for an experimental powder XRD patterns. Photomicrographs were taken during isothermal dehydration combined with calculated BFDH morphology to help distinguish concomitant polymorphs [156], to understand the faces, and directions of solvent loss [157], while atomic force microscopy supported with BFDH calculations was used to follow phase changes *in situ* [158]. Analysis of intermolecular interactions over molecular surfaces allows the rationalization of concomitant polymorphism of substances where single crystals of two crystal forms share faces of their crystals [159]. The case of intergrowth polymorphism [160,161] referred to the existence of two solid forms within one single crystal, which is a special case of such concomitant polymorphism. On the contrary, absence of similar surfaces on dominant phases of 1:1 and 3:2 co-crystals of *p*-toulenesulfonamide and triphenylphosphine oxide allowed Croker, et al. to conclude that solvent mediated phase transition in this system occurs through dissolving of one phase and re-crystallization of the other [162]. Analysis of polymorphs, solvates, and hydrates of trospium chloride containing structures demonstrated that these compounds can be divided into three main structural groups, and the predicted and experimental crystal morphology of various forms within one structural group was found to be very similar [113]. The BFDH morphology tool was applied to rationalize silicon oil induced spontaneous phase transition in ethynyl-substituted benzamides [163] and unusual mechanical response from a crystal undergoing topochemical dimerization [164].

More complex applications of BFDH morphology tool include synthesis of desired solid forms or prediction of desired properties for previously published compounds. It was shown that self-assembled monolayers of rigid biphenyl thiols [165–167] or silioxane-based monolayers [168,169] can be used as templates to exclude concomitant polymorphysm (Figure 6a) or to obtain a metastable polymorph. First, analysis of crystal faces of the desired solid form was carried out to reveal functional groups on one of the main faces; then, the monolayer containing functional groups are able to form stable synthons with the above functional groups, which was used during crystallization (Figure 6b). Not only predesigned monolayers but also polymers [170,171] and additives (Figure 6c) [172,173] can be used as templates for crystallization. For example, β-1,4-saccharides act as templates to produce the metastable form III of paracetamol at crystallization from melt readily forming H-bonds with the (010) surface of form III [172], while the presence of sulfamides in solution promotes crystallization of γ-pyrazinamide from aqueous solutions [173]. Pyrazineamide usually nucleates from solutions in its' α form that contains molecular H-bonded dimers; amide of pyrazineamides probably forms heterodimers with sulfamide groups of templates on one of the crystal surfaces and then forms N-H...N connected chains with neighboring molecules. These chains are found only in γ-polymorph. Patel, Nguyen & Chadwick [171] not only used polymers to promote heterogeneous nucleation but also suggested to use *H-bonding propensities tool* to rank polymer surfaces towards heterogeneous nucleation of benzocaine and 1,1'-bi-2-naphthol.

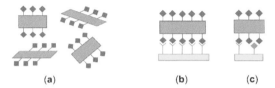

(a)   (b)   (c)

Figure 6. Schematic representation of (**a**) concomitant polymorphism, (**b**) crystallization on self-assembled monolayers, (**c**) crystallization on additives.

Applications of some anisotropic (needle, plate) crystals are related with anisotropy of chemical bonds within such solids. Anisotropic crystals with predominated faces formed by hydrophobic groups can be used as superhydrophobic porous materials [174]. Analysis of mechanical stimulus applied along three axes of unit cells to some layered energetic materials allows us to conclude that these explosives can convert kinetic energy into layer sliding to prevent the formation of hot spots [175]. This should also be the rationale for why high-performance insensitive energetic materials can be used as desensitizers versus mechanical stimuli. Vice versa, a corrugated molecular structure of celecoxib without any slip-planes and numerous weak interactions in orthogonal directions were suggested to be the reasons for exceptionally high elasticity of its needle singe crystals [176]. Flat naphthalene diimine derivatives were found to have comparable crystal packing governed by π...π stacking, similar crystal morphology, but exhibited various mechanical flexibility attributed to close packing or interlock of their terminal alkyl chains [177]. Investigation of sulfa drug crystals [178], co-crystals of vanillin isomers [179], and amino acids [180] demonstrated that H-bonded layered structures with orthogonal distribution of strong and weak interactions attain the feasibility of cleaving a crystal along a given crystallographic plane parallel with these robust synthons.Particularly, single crystals with molecular surfaces formed by hydrophobic interactions can be applied as clean surfaces for molecular beam epitaxy. The presence of slip planes is also thought to be associated with low elastic recovery upon compression, and greater plasticity [181,182]. Crystal morphology prediction was applied to discover new gelators by focusing on scaffolds with predicted high aspect ratio crystals [183].

Thus, analysis of crystal morphology and functional groups on main crystal faces cover many areas of crystal design and are important for many areas of industry. The tools implemented to the Mercury package can be helpful in these areas, but their application seems to be limited with anisotropic crystals or solids faceted with different functional groups. In other cases, this approach can be improved by DFT calculations of the interaction effect between the growing faces and the solvent molecules, or molecular electrostatic potential on crystal faces.

5. Knowlede-Based Prediction of Supramolecular Associates and Solid Forms

5.1. Polymorph Assessment

Understanding and prediction of polymorphism—the ability of a solid material to exist in multiple crystal forms known as polymorphs—is vital for the industry as polymorphs of drugs, explosives, and pigments exhibit different properties [55]. It can be associated with various molecular conformations of flexible molecules, competing intermolecular interactions of polytopic molecules, various packing of molecules or supramolecular synthons, et cetera. While some paths to control polymorphism are described in Section 4.1, here, the approach to estimate the likelihoods of H-bonded polymorphism is described. It is of special interest for the analysis of compounds containing few donors (D) and acceptors (A) of H-bonding, since competition of donors or acceptors to take part in a hydrogen bond, and restrictions to form all highly likely bonds simultaneously may cause the appearance of polymorphs [184]. Taking into account that various polymorphs exhibit different properties including solubility, bioavailiability, tabletability, et cetera, the pharmaceutical industry has prompted the development of computational tools that are able to predict the likelihoods of H-bonded polymorphs. Thus, a methodology has been developed [185] to estimate the likelihood of H-bond formation for each pair of donors and acceptors in a molecule taking the environment of the D and A groups into account.

The *H-bond propensities tool* uses only a 2D molecular formula (Figure 7) and involves four stages: Data sampling, model fitting, model validation, and target assessment. For each functional group, it is either automatically assigned from the 2D formula, taken from the functional group library, or sketched manually by using a fitting data set, which is generated by loading in an existing set of structural data, or from the CSD. The H-bonds within the fitting data set are identified to collate statistics and descriptors of H-bonding formation. The logistic analysis is performed to generate a statistical model to determine the likelihood of H-bond formation; the model should be analyzed to include for the final model only significant variables and be good enough to proceed (area under ROC curve (receiver operating curve) above 0.8 indicates good discrimination). The results are ranked by propensity to allow inspection of the most and the least likely D-A pairs. Based on this knowledge-based approach, a structure that adopts the highest propensity H-bonds displays a low likelihood of appearance of a more stable polymorph, while for solid forms with lower propensities of H-bonds, a high risk of polymorphism caused by H-bonding is expected. For example, the most likely interaction for a "molecule" depicted in Figure 7 includes a donor atom of Group 3 and an acceptor atom of Group 1. However, in solid {Refcode} the second likely hydrogen bond between groups 1 and 2 is observed, thus, indicating possibility of appearance of another polymorph with the most likely hydrogen bond.

Figure 7. Flowchart for the methodology to predict the likelihood of hydrogen bond formation.

The tool has proved to be highly valuable in estimating the relative stability of known crystal forms [186]. Particularly, the kinetically more favored **Form I** of ritonavir displays statistically unlikely hydroxy-thiazoyl and ureido-ureido interactions but exhibits more favorable conformation of the carbamate moiety, while **Form II** realizes stronger hydrogen bonds as estimated in Ref. [186]. Two X-rayed polymorphs of 4-aminobenzoic acid realize different H-bonding, and the more stable form realizes more likely sets of H-bonding [187]. The approach was also successfully used to rationalize polymorphism for N2-(indol-3-acetyl)-L-aspargin [186], lamotrigine [188], and crizotinib [189,190] (see Figure 8 for chemical formulas of these compounds). Not only strong donors and acceptors of

H-bonding can be used for analysis; a competition of hydroxide O-H and ethynyl C−C≡C−H groups to form hydrogen bonds [191] and the role of CHCl₃ and CH₂Cl₂ molecules in crystal packing [192] were demonstrated using this tool.

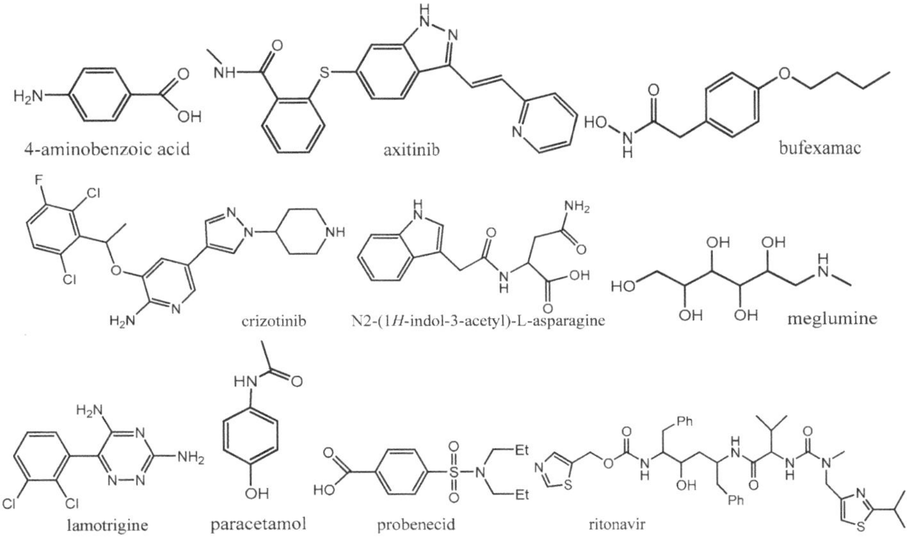

Figure 8. Schematic representation of compounds analyzed using H-bonding propensities tool.

Known forms of paracetamol both exhibit H-bonding between OH groups, that is a less likely donor to form a bond than an amide fragment. It was assumed that there was a conflict between a good donor group to form a bond disfavoring a poor acceptor to form bonding, which could be overcome by co-crystal formation with molecules containing the same D and A groups [187]. Indeed, co-crystals of paracetamol with diamino- or bis(4-pyridyl)-containing co-formers contain molecules of paracetamol connected by more likely amide/hydroxo or amide/amide pairs of interactions [193]. A good correlation between H-bonding charge analysis and H-bond propensities was also demonstrated for crizotinib [189], and heterocycle-1-carbohydrazoneamides [194].

Note, that the high probability for exhibiting polymorphism does not necessarily allow one to obtain novel polymorphs and that different polymorphs can realize the same H-bonds [195]. For example, probenecid was predicted to realize various H-bonded polymorphs, and differential scanning calorimetry indeed revealed three polymorphs, but XRD experiments showed that all of them realized the same hydrogen bonding [196]. For bufexamac, the propensity tool suggests the possible existence of three crystal forms, while Nauha and Bernstein found two polymorphs with similar H-bonds [197]. Two experimentally observed forms of meglumine realize slightly different H-bonding, which, however, are absent on putative structure landscape as these contain highly unlikely bifurcate acceptor OH groups [197]. For axitinib [189], five anhydrous polymorphs realize the most likely H-bonding, thus, the approach can not distinguish the most stable among them.

To sum up, the propensity tool neither predicts polymorphism, nor guarantees that all sets of theoretically possible H-bonding combinations can be obtained or guidelines can be given on how to obtain any of these polymorphs. It does not describe inclinations of a molecule to form concomitant polymorphs, conformational polymorphs, or packing polymorphs, or solid forms with Z′ >1. Instead, it indicates the possibility of solid forms to organize various H-bonded architectures and provides some guidelines on the amount of affords that one can spend in the experimental search for various crystal forms. Abramov also notes such a limitation of the approach as incapability to distinguish more stable polymorph among crystal forms with similar H-bonded networks, which however can be overcome with additional charge-density analysis of known polymorphs [189]. At the same time, its application is not limited by monomolecular systems. Note, that any desired functional groups, additional co-formers, and solvent molecules can be included in the statistical model. Thus, although this method is not able to predict the ratio of components in a co-crystal, it can be applied for co-crystal design.

5.2. Co-Crystals Design

Co-crystals are solids that consist of two and more components (co-formers) that form a unique crystalline structure having unique properties. Thus, co-crystallization is an approach to optimize the physical properties of solid materials. For example, pharmaceutical co-crystals can be obtained to modulate dissolution rate and physical stability of drugs [198,199]. Schultheiss and Neuman reported [200] the effect of coformers on melting points, stability towards humidity, solubility, bioavailiability, and some other properties of pharmaceutical co-crystals. Karki, et al. [201] demonstrated that the tabletability of pharmaceuticals can be tuned up on the example of paracetamol co-crystals with various co-formers. The dependence of mechanical properties of acid...amid based co-crystals was studied by Saha and Desiraju [202]. The potential of co-crystallization for tuning the properties of the energetic materials [175,203,204], optical materials [205,206], and for the food industry [207] have also been demonstrated in recent papers. In contrast with synthetic routes focused on the synthesis of covalent derivatives containing functional groups that affect the desired property, co-crystallization became of interest as a 'greener' process, frequently free of toxic ingredients and by-products. Pharmaceutical compounds in this context become of particular interest as these molecules contain functional groups that can be involved in molecular recognition of biomolecules [208]. The main task of crystal chemistry in this field is to estimate which co-formers, if any, will form a co-crystal with a given molecule prior to screening and investigation of properties of the solid obtained.

One group of methods available are quantum chemical calculations from simple energy minimization to full structure prediction [209]. Another group of methods is based on data knowledge about the likely homo and heterosynthons. Since the works of Etter [210], it is known that (i) all good donors and acceptors tend to take part in hydrogen bonding, (ii) the strongest donors tend to interact with the strongest acceptors. In this context, it becomes essential for a researcher to estimate which donors and acceptors are "good" and "the strongest". The *H-bonding propensities* tool allows ranking of various donors and acceptors without any quantum-chemical calculations (Figure 9).

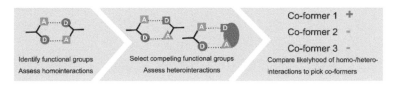

Figure 9. Flowchart for co-crystal design strategies based on synthon competition.

H-bonding propensities tool for uncharged and charged molecules of pyrimethamine and dicarboxylic acids, as well as halogenated aromatic compounds (Figure 10a), was successful in predicting formation and non-formation of adducts [211]. A similar approach was later applied to pyrimethamine with some other drug molecules also taking solvent molecules into account [212]. Note, that proton transfer accompanied with transformation of a co-crystal to a salt was rationalized in terms of their ΔpK_a values [213,214], where $\Delta pK_a = pK_a$[protonated base] $- pK_a$[acid] >2 or 3 was found for salts, negative ΔpK_a was characteristic of co-crystals, while an intermediate situation may result in both depending on stable supramolecular synthons [215]. Hydrogen-bond propensities gave similar trends as calculated bond energies for possible synthons in thiazole amides (Figure 10b), and co-crystallization of six thiazole amides with 20 different carboxylic acids demonstrated effectiveness of complimentary energy-based and data–knowledge predictions for the prediction of likely synthons [216].

The knowledge-based approach to co-crystal design was successfully applied to select pyrazinoids and pyridinoids as prospective co-formers for a diuretic drug hydrochlorothiazide [217] to tune its solubility. The amide-pseudoamide synthon was shown to be more stable than the formation of two dimers of the theophylline molecule, that allowed to explain the formation of theophylline: Amide co-crystals [218]. At the same time, theophylline co-crystals with fluorobenzoic acids demonstrate that it is still difficult to predict co-crystallization of small rigid molecules using this method [219]. Similarly, preferable co-crystallization of only one of three tautomers of 2-amino-6-methyl-1,4-dihydropyrimidin-4-one with carboxylic acids was explained using this approach [220]. Potential of co-crystal design for molecules with limited H-bonding functionalities was demonstrated on the example of propyphenazone [221]: NH_2, OH, and CO_2H functionalities were found to be the most likely groups to interact with O=C group, and eight novel co-crystals were obtained using co-formers containing OH or/and CO_2H groups. Dicarboxylic acids [222] or anions [223,224] can be applied to co-crystallize two molecules, which do not form co-crystals.

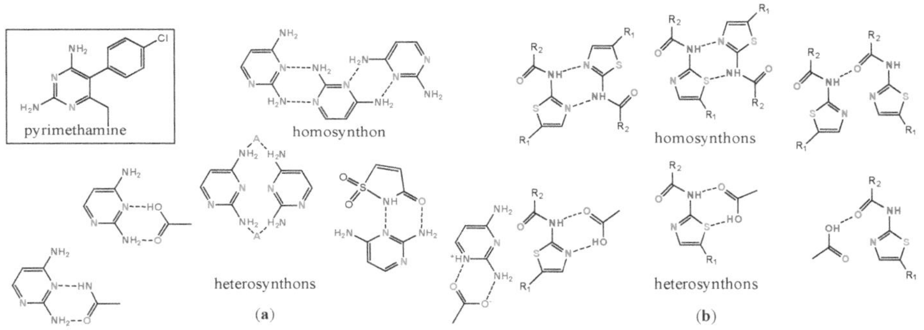

Figure 10. Examples of homo/heterosynthons analyzed for (**a**) pyrimethamine [211,212], and (**b**) thiazole-amides [216].

Note, that the analysis of possible heterosynthons is not limited to co-crystals. It can be utilized to investigate inclusion compounds [225] and mixtures of polymers and solutes [171]. Besides, the ranking of heterosynthons on their relative strength can not only help in estimating the most likely binary co-crystals. This method has great potential in the synthesis of multicomponent co-crystals based on polytopic co-formers. Thus, H-bonded heterosynthons are able to interact with the third co-former through halogen bonds [226–228]. π-Stacking in conjunction with hydrogen bonding was used for synthesis of ternary co-crystals [229,230]. Combination of H-bonded synthons and stacking interactions between donor and acceptor planar molecules allowed Desiraju and co-workers to synthesize quaternary and even quintinary co-crystals [231,232]. Partial substitution of some co-formers in quaternary systems with their shape-size analogs even allowed for obtaining six component solids [233].

Although some examples of co-crystals design based on H-bonding propensities tool are given above, successful prediction of inclination of some molecules to form co-crystals obviously needs analysis of some other molecular descriptors. First, the possibility of a solvent to take part in H-bonding should be taken into account [211,212,234]. A systematic study of co-crystallization of paracetamol with H-bond acceptors [235] and donors [236] demonstrated that paracetamol molecules in these crystals are linked via either OH···O=C or NH···O=C interactions, depending on the presence or absence of substituent groups on the molecule of the second co-crystal former. A machine learning algorithm trained out on a set of paracetamol co-crystal experiments using more than 190 molecular descriptors for each co-former allowed to predict 9 of the 13 experimentally obtained co-crystals within the top 11 suggestions [237]. Unfortunately, this method requires a large amount of experimental work to train the model, and the model obtained should be applied only to molecules with similar molecular descriptors as those for an attested molecule. Fabian has proposed a methodology that involves using size and shape complementarity as the primary driver for co-crystal formation [238]. Analysis of 131 molecular descriptors for 710 co-crystal partners suggested that co-crystals were more likely to form between molecules of a similar size and shape and with a similar polarity of co-crystal formers, thus both these descriptors were recently included to the *Co-Crystal Design tool*.

Effectiveness of such modification can be demonstrated on the example of 1,2,4-thiadiazole derivative co-crystallization with gallic and vanillic acids [239]. While *H-bonding propensities tool* gives almost equal probability of occurrence of homo- and heterosynthons, molecular complementarity tool indicates that co-crystals in these systems should form; and these co-crystals were experimentally obtained. Karki, et al. [240] demonstrated that synthon analysis combined with Fabian's methodology was effective in prediction of possible co-formers for artemisinin. A series of co-crystals of sulfamethoxazole [241] and leflunomide [242] were synthesized using this approach. Prediction of possible H-bonded motifs between tyraminium cations and violurate anions also successfully predicted many of the bimolecular synthons experimentally observed in tyraminium violurate polymorphs and hydrates, but also demonstrated that none of the trimolecular synthons were predicted [243]. Note, that tri- and tetra-molecular synthons are not something unusual in co-crystals (see, for example, tetramolecular associates found in thiazole-amides, Figure 10), thus any predictions should also include such polymolecular associates into account.

To sum up, *H-bonding Propensities* and *Co-Crystal Design* tools allow one to simplify co-crystal screening. CSD analysis helps to select complementary functional groups to form heterosynthons more likely to form than homosynthons. Presence of false positives in experimental screening then may be accounted for other factors. For example, although hydrogen bonds are thought to be stronger than halogen bonding, one can undergo a competitive co-crystallization between H-bonded and halogen bonded synthons based on the polarity of a solvent used [244]. It was shown that in polar solvents 1,2-bis(pyrid-4-yl)ethane forms co-crystals via halogen bonds, and in nonpolar - via H-bonds. Of more concern with regards to co-former screening is the possibility of false negatives. If the results of this type of analysis are used to narrow down the number of screening experiments performed, any possible co-crystals that are incorrectly marked as unlikely to form would be missed. Besides quantum calculations of molecular dimers or logistic models trained on large datasets of co-crystals, development of shape-size molecular complementarity tools for the prediction of possible co-formers seem to be very promising. Note, that the CSD-Crossminer package developed for sophisticated search of analogues of drugs is now available (Section 3.4). It utilizes the idea that molecules with similar disposition of functional groups and hydrophobic fragments may act similarly with proteins and replace each other. The idea of substitution of one of the co-crystal formers with its size-shape analog can be realized using this program.

5.3. Full Interaction Maps

The *Full Interaction Maps* (FIM) tool [245] implemented within the Mercury package visualizes the likelihood of a synthon appearance and corresponding geometry variation between functional groups of a molecule under consideration, and a probe functional group. For each functional group of a molecule, 3D scatterplots of CSD contact searches with a chosen probe, and the functional groups are generated and converted into scaled density maps. These density maps are then combined for the whole molecule, taking the environmental effects of combinative factors and steric exclusion to account. The 2019 version of the CSD-Enterprise contains (i) RNH_3, uncharged and charged NH nitrogen atoms as probe functional groups of donors of hydrogen bonds, (ii) various oxygen atoms (of a carbonyl and alcohol group or a water molecules) as a probe of hydrogen bond acceptors, (iii) methyl and aromatic carbon atoms as a probe of stacking interactions. The CSD-Enterprise version of 2018 allowed the use of C-I and C-Br interactions as a probe of (iv) halogen interactions; and the next year these interactions were extended with C-F and C-Cl probes, to evaluate differences between various halogen atoms. For each functional group, the FIM distribution is similar with *IsoStar* scatterplots of corresponding functional groups, but the FIM for a whole molecule reflects the fact that some of functional groups are better donors and acceptors of, for example, H-bonding, than the others. On these maps, this difference is expressed in the color (and the coordinates of the most likely positions of the donor and acceptor sites can be additionally calculated and depicted), while corresponding quantitative values can be estimated using the *H-bonding Propensities* tool (Section 5.1). Besides, these maps are very sensitive to the steric exclusion from other molecular species and would differ for a molecule in optimized and experimental geometries, and those obtained with CSD-Conformer Generator.

The FIMs probed with H-bond donors (blue) and acceptors (red) and supramolecular synthons of the *Pbcn* polymorph of chalcone {BZYACO01}and 5-(4-bromophenyl)-1-phenylpent-2-en-4-yn-1-one {YILNAK} are depicted in Figure 11. For both molecules' blue regions near the ketone oxygen atom reflect its inclination to form bifurcate bonding due to the presence of two lone pairs on the oxygen atoms. The most expected positions of acceptors of hydrogen bonding for the chalcone are situated on the opposite sites of the molecule, and correlate well with observed supramolecular synthons in its polymorphs—a dimer depicted in Figure 11, a {BZYACO01}and a head-to-tail chain {BZYACO04} similar with that of 5-(4-bromophenyl)-1-phenylpent-2-en-4-yn-1-one. Elongation of the π-conjugated chain with a triple bond makes the appearance of such chains more abundant than dimer occurrence (Figure 11b), that was experimentally confirmed for a series of 1,5-diarylpent-2-en-4-yn-1-ones [246,247].

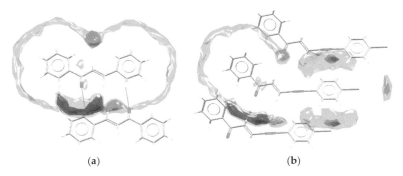

(a)   (b)

Figure 11. Interaction maps and supramolecular synthons in the crystal structures of (**a**) chalcone {BZYACO01}, (**b**) (E)- 5-(4-bromophenyl)-1-phenylpent-2-en-4-yn-1-one {YILNAK}.

This tool has been applied mainly for the analysis of strong hydrogen bonds [54,170,190,194,248–251] in accord with the first test functional groups suggested within this tool, and the first paper published to describe it. Mutual disposition of hot spots can be used to find functional groups of the same molecule [245], co-formers [243,249,250], active sites in a binding pocket of a macromolecule [252], or surface inhibitors [170], which match a given pattern of interaction preferences. Analysis of polymorphs of some drugs demonstrated that disposition of H-bond donors and acceptors close to the hotspots are indicative for the more stable polymorph even if a molecule realizes less likely conformation [190]. However, in the case of polymorphism induced by weak intermolecular interaction reorganization [253] or high-pressure [254], it has lower predictive ability. This can easily be understood if we keep in mind that data for the FIMs are plotted based on structural information derived at atmospheric pressure.

At the same time, the test carbonyl oxygen and uncharged NH groups can be successfully applied to reveal positions of acceptors and donor of C-H...O [170,255–257] and C-H...N [258] bonding. It was demonstrated in Refs. [255,257] that the FIMs can be applied to estimate likely C-H...O bonded motifs for chalcones, polyenones, pentenynones and cyclic ketones with vinylacetylene fragments. At the same time, it was demonstrated that C=O and C-Br groups compete with each other for the most acidic hydrogen atoms, thus that FIM predictions of synthons based on C-H...Br bonding became less reliable at the presence of C-H...O=C bonding [257]. At the absence of carbonyl groups the FIMs for the N-salicilidenanylines probed with C-I and C-F groups successfully predict all of the C-I...N and many of the C-F...H-C interactions in co-crystals of N-salicilidenanylines with perhalogenated co-formers [259]. Mugheirbi and Tajber analyzed the FIM hotspots around the itraconazole molecule to understand the molecular environment in the mesophase [260]. The itraconazole lacks any donors of strong H-donors but has a number of competing acceptors of H-bonding; thus, analysis of the FIMs and FTIR spectra allowed them to reveal the most ordered of mesophases, and to propose that the greatest mobility of this molecule is associated with movement of the triazoline ring. Analysis of FIMs in two polymorphs of the dinuclear Co(II)-Shiff base complex revealed the most acidic H(C) atoms (similar in both polymorphs), which take part either in C-H...O bonding in the triclinic polymorph, or in the C-H...π bonding in the monoclinic polymorph, or even to not take part in any prominent intermolecular bonding [256].

Note, that unsatisfied strong acceptors observed in a crystal structure solved from powder XRD data may be indicative of missed water/solvent [261,262]. In this case coordinates of the most expected position of a water molecule can be used in the refinement. Worth noting, that *IsoStar* and FIMs deal only with intermolecular interactions but a similar idea could be used for investigation of metal-ligand bonding in polytopic ligands. This could be helpful for understanding the factors that govern linkage isomerism, and to estimate the most likely coordination mode of the most widespread ligands.

5.4. CSD-Conformer Generator

Representation of a molecule in the three-dimensional space finds numerous applications in structure solution from powder diffraction data in real space, crystal structure prediction including the formation of co-crystals, protein–ligand docking, and others. Thus, the *CSD-Conformer Generator* tool included in the CSD-Enterprise as an approach to a fast generation of plausible molecular conformations by using geometric distributions derived from the CSD. First, for the input 3D molecular model with all hydrogen atoms present, all bond lengths and angles are minimized based on corresponding average values. Then, the molecule is partitioned into rotamers, and rotamer libraries and ring template libraries are used to generate a conformer tree with unforbidden and preferred rotatable bond geometries and ring geometries. A final set of conformers is clustered according to conformer similarity. Each conformer is locally optimized in torsion space. It was demonstrated that this tool reproduced well conformations of a number of molecules observed in the CSD and the Protein Data Bank [263,264]. Some discrepancies between the predicted and experimental geometries occurred for the ligand: Macromolecule complexes in unusual conformational space, rare rotamer examples, uncertain bond types, and some other cases, nevertheless, theoretical configuration space represents well experimental data [265,266] or configurations obtained by the Molecular Operating Environment's Low Mode Molecular Dynamics module [267]. It was demonstrated that the CSD-Conformer Generator combined with ab initio [268] or DFTB3-D3 [269] calculations can be used for crystal structure prediction of some flexible pharmaceuticals. Such combination reproduced well the molecular geometry and crystal parameters, although did not provide sufficiently accurate energy ranking.

6. Hydrate/Solvate Analyser

As it was described above, a solvent can affect crystal morphology and sometimes determines the polymorph, but it also can be built into the crystal structure to form a hydrate or solvate. Since water is nature's solvent, non-toxic and widespread, the scientific community is interested in understanding water assembles in liquids and solids. In particular, medium-sized and large water clusters are important for biology, since they act as surrounding and solvating solutes for biologically active molecules, fill discrete voids and channels in molecular and supramolecular assemblies including their reactive sites and interpenetrate into the interfacial region of hydrophobic surfaces [270–273]. Besides, water associates are involved in dynamic processes such as proton transport, material or molecular folding, de/resolution, and others [274,275], and even affect mechanical properties of single crystals [276]. The behavior of liquid and solid water including ice, clathrates, and ice-like systems are determined by the disorder of the hydrogen atoms, and the binding energy of different configurations as well as some other properties can also differ [277–279]. At last, but not least different solvents may stabilize different forms of a molecule (the neutral or zwitterionic forms) [280,281]. In other words, both (i) connectivity of a solvent molecule in a crystal, inclination to form a given associate, and (ii) dimensionality and unit cell volume that goes to solvents are of interest for biology, crystallography, and material chemistry. All corresponding algorithms are realized within the *Hydrate* and *Solvate Analyser* tools of the CSD-Material module. The *Hydrate Analyser* provides information about the water H-bond geometry and motifs (one of any of 10 most common motifs described by Ref. [282]) detected, information on the volume occupied with water molecules, and display the water space and water interaction maps. The *Solvate Analyser* provides similar information about simple and mixed solvates, co-formers and ions.

6.1. Analysis of the Local Connectivity of a Solvent

Analysis of solvate connectivity and the most abundant motifs (if any) formed by solvent molecules is similar to that described above for the investigation of other synthons and supramolecular associates. The practical meaning of results obtained for material chemistry, biochemistry, and the pharmaceutical industry are based on the fact that similar hydrate architectures were found in crystals of both inorganic, organic, coordination, and even macromolecular compounds [40,41]. H-bonding between a complex and an outer sphere solvent can lead to additional quenching in luminescent materials that are unfavorable for the task-specific design of optical materials [283,284]. Comparison of clathrate, turbolato-clathrate, and non-clathrate hydrates revealed that water associates in the latter can be regarded as fragments of clathrate hydrates [285]. Analysis of hydrate architectures in crystal structures of known kosmotropic and chaotropic agents revealed that the kosmotropes tend to take part in H-bonding with hydrates, while chaotropes in crystals tend to from clathrate hydrate-like structures [286]. Kosmotropic agents are those able, by the ordering of the water structure in solution, to enhance intermolecular interactions within protein molecules, thus preventing denaturation, while chaotropes act oppositely. Hence, the property of a molecule to crystallize with water giving full enclathration was directly associated with the ability to act as chaotrope.

Most frequently water serves as a donor of two hydrogen bonds and an acceptor of one hydrogen bond, although three other motifs are also relatively common (Figure 12a [282,287,288]). The dual nature of this molecule in respect to H-bonding allowed proposing imbalance in the number of donor and acceptor groups of a polytopic molecule to be the reason for form hydrate appearance [289]. Although, it is in accordance with Etter's rule, which states that "all good proton donors and acceptors are used in hydrogen bonding" [210], Infantes, Fábián, and Motherwelldemonstrated that the imbalance of donor and acceptor groups, in fact, does not affect inclination of a molecule to form hydrates [288]. Instead, it more readily interacts with unsatisfied acceptors of H-bonding, especially $R_2PO_2^-$, Cl^-, $C-NH_3^+$ groups, acts as a bridge between unsatisfied donors and acceptors, and occupies the free volume, especially for chiral molecules [290]. The strong imbalance between the number of donors and acceptor groups between tetrasulfonate-functionalized rigid anions and planar polyamino-containing cations indeed afforded their crystallization as hydrates, where water molecules act both as bridges between cations and anions, and clusters incorporated within cavities and channels of H-bonded networks [251]. Nevertheless, typically it is still hardly possible to predict if a compound will form a hydrate or not. Zaworotko and coworkers revealed that not only molecules with unsatisfied donors but also acceptors of H-bonding also readily form hydrates, but the reason of CSD statistics, in this case, might underestimate inclination of a molecule to form a hydrate, probably, because the most effective pathways to hydrate formation were slurrying in water and exposure to humidity [291]. The role of MeOH [292], DMSO [293,294], and $CHCl_3$ and CH_2Cl_2[192] solvates in crystals was also investigated.

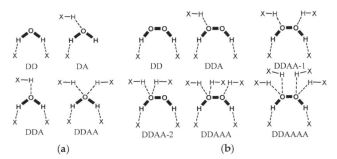

Figure 12. H-bonding patterns in (**a**) organic hydrates [288] and (**b**) organic peroxosolvates [116] (X denotes any donor or acceptor atom).

Comparison of motifs formed by water molecules in organic hydrates and by H_2O_2 in organic peroxosolvates gave a clue to the understanding of their isomorphysm [116,295]. It was found that H_2O_2 always forms two H-bonds as a proton donor, and up to four bonds as an acceptor of H-bonding (Figure 12b). Thus, only three of the peroxosolvate H-bonding patterns are similar to those observed in crystals of hydrates. As isomorphous substitution may occur only in crystals with similar H-bonding motifs, peroxosolvates with DD, DDA, and DDAA motifs connectivity require high H_2O_2 concentrations for synthesis, are sensitive to humidity, and as a result are of limited practical meaning. Instead, peroxosolvates involved in at least five hydrogen bonds and strong H-bonding with acceptors are not inclined to take part in isomorphous substitution with hydrates.

Not only water connectivity, but also the main motifs formed by water molecules, or supramolecular architectures including water molecules as well as suggested nomenclature were described [67]. Combined together, these data have practical meaning for prediction of coordinates of missing water molecules for solving and refinement of crystal structures from powder X-Ray diffraction data [261], as well as for molecular docking with GOLD (Genetic Optimization for Ligand Docking), part of the CSD-Discovery module.

6.2. Analysis of Solvent Associates

Thus, besides analysis and classification of H-bonded motifs, visualization of associates, and calculation of the unit cell volume occupied with water/solvent molecules (both coordinated and not) is available via *Hydrate/Solvate Analyser* tool. On Figure 13, a DMSO and water space in the structure of bosutinib DMSO solvate trihydrate {ABECES} are shown in red and blue, respectively. It is clearly seen that both solvents form isolated clusters, while in the structure of tetrabutylammonium fluoride clathrate hydrate {CIPRAV} H-bonded water molecules form a 3D architecture with tetrabutylammomium cations situated in pores of this framework (Figure 13b). Visualization is similar to void analysis, but there is no need to remove any molecules to carry out calculations. Moreover, the *Solvate analyser* tool can be used to identify not only solvents, but also co-formers and ions, disordered fragments, or even hydrophobic/hydrophilic fragmentset cetera, as well as to visualize their packing and check H-bonding involving the identified components.

Analysis of the voids within previously published copper-containing coordination polymers allowed Inokuma et al. to find a crystalline sponge with desired pore size and affinity to guest molecules and demonstrated that a huge number of previously published coordination polymers can be used to capture guest molecules [296]. Anisotropic motifs formed by solvent molecules affect mechanical properties of compounds, and dehydration mechanism. For example, presence of infinite H-bonded water chains is typically associated with the reversible dehydration process and similarity between crystal structures of the hydrated and anhydrous forms of a solid, like caffeine [297], carbamazepin [104], aspartame [298], and Shiff-base Ni(II) complex [299]. Such channels can be not only dehydrated but also substituted with other solvents [300,301]. The solid dehydration with prominent atomic movement traced with single-crystal or powder X-Ray diffraction can shed light on a more complex mechanism of water loss [302–306]. Various mobility of the water clusters incorporated within the pores and H-bonding waters that structure these pores was demonstrated in Refs. [251,307]. Liu et al. demonstrated that anhydrous uric acid and its' dihydrate exhibited various mechanical properties of their single crystals despite the similarity of layered motifs formed by the acid [276]. 2D water associates situated between acid layers perpendicular with single crystal main faces break at indentation with dehydration at crystal surface and make this solid softer than the anhydrous form.

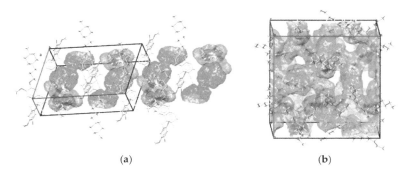

(a)   (b)

Figure 13. The structure of (**a**) bosutinib dimethyl sulfoxide (DMSO) solvate trihydrate {ABECES} and (**b**) tetrabutylammonium fluoride clathrate hydrate {CIPRAV} with solvate analyser display of DMSO space shown in red and water space in light blue.

7. Additional Software Compatible with the CSD-Materials Analysis of Structure–Property Networks

7.1. CSD Python API

The *CSD Python API* is now part of the CSD-Enterprise automatically installed as part of the CSD installation. This allows one to run supplied or user-written Python scripts for the loaded structure or a set of structures. CCDC Python-Built-In scripts include at the moment:

- Analysis (generation of conformers and calculation of their RMSD (*conformes_similarity.py*) or of all torsion angles (*generate_conformers.py*) for the loaded structure).
- External *(load_in_conquest.py*– loading the current structure into the ConQuest sketch window).
- Reports (generating a report in *.html format containing crystallographic details, molecular geometry, and intermolecular hydrogen and halogen bonding (*crystal_structure_report.py*), or molecular geometry, including Mogul geometry analysis (*molecular_geometry_report.py*), or crystal packing descriptors (*quick_packing_check.py*), or basic chemical, crystallographic and publication information about the loaded structure (*structure-simple_report.py*)).
- Searches within the CSD for entries relevant to the specified chemical name or synonym (*chemical_name_search.py*) or having similar molecular geometry (*molecular_similarity_search.py*).
- One can additionally download from the CSD-Python API Forum [308] the following scripts:
- To generate molecular formula and weight (*welcome-and-weight.py*), Crystal14 input files (*crystal_inputs.py*) or Packing Similarity dendrogram (*Packing_Similarity_Dendrogram.py*);
- To send a Mol2 or CIF files to an external application (*send_mol2_to_notepad+.py*);
- To generate diagrams for all molecules in a structure file (*diagram_to_file.py*);
- To find covalently bonded clusters within a structure (*dimensionality.py*);
- To perform void calculation in a crystal (*void_calc.py*);
- To merge GOLD docking results (*gold_merge.py*);
- To filter molecular conformers with unusual torsion angles (*conformer_filter_density.py*);

- To extract from the CSD unique rings of a specified size (*ring_type_count.py*);
- To compare protein bound ligand with CSD-Generated conformers (*find_binding_conformation.py*).

Some other examples of Python scripts applied for investigation of crystal structures include removing solvent molecules from a MOF subset extracted from the CSD to perform analysis of geometry and properties of nondisordered porous compounds [309]; authors realized the possibility to remove by request both uncoordinated and coordinated solvent molecules from coordination polymers. Dolinar et al. wrote a Python script to perform the CSD search of chiral monomolecular compounds [310]. Bryant, Maloney, and Sykes investigated a number of polymorphs and co-crystals with previously reported tabletabilities to analyze how different factors govern tabletability of organic crystals, and to suggest a programmatic method to calculate corresponding crystal descriptors [311]. Their investigation of the most likely slip planes, their interpenetration on connection by weaker interactions, presence of other slip planes, and automatic assessment of H-bond dimensionalities revealed that the degree of separation, followed by the presence of H-bonds between the layers, and finally d-spacing between potential slip planesare the most important descriptors affecting the tabletability. The corresponding CSD Python API protocol that can be applied for the prediction of mechanical properties of organic crystals was also published [311]. Miklitz and Jelfs reported a Python script for structural analysis of porous materials (organic and coordination cages) that estimates the cavity diameter, number of windows, and their diameter [312]. It uses Cartesian coordinates and atom types for input, thus, can be in principle applied also to selected fragments of framework materials. Besides codes to extract and analyze data stored within the CSD, for text mining of selected properties within a large set of papers was also published [313]. The algorithm uses manuscript html files as input for extraction of surface areas and pore volumes of metal–organic frameworks with at least 73% accuracy and can be used for investigation of structure–property networks.

Thus, the Python API gives a convenient path to combine and modify the CSD requests with CSD-Materials, CSD-Discovery tools or external software for ones' purposes. The next Section gives some examples of the combination of the CSD-Materials tools with external crystallographic software.

7.2. Combination of Various Tools and Algorythms

Some combinations of the CSD-Materials tools with external software were mentioned in previous Sections, for example, the combination of the *CSD Conformer Generator* with periodic calculations to predict crystal structures, or that of the *CSD Co-Crystal Former* and the *H-bond propensities* with DFT calculations to better understand competing for intermolecular interactions. Let us mention also some other examples, where the *CSD-Materials* tools were used in unusual way or as the tool to solve only part of a scientific problem.

Crystal packing of organic single-component compounds with Z' = 1 were compared to evaluate the effect of chemical transformation on crystal packing and isostructurality using a combination of the CSD Python API scripts, external software, and the *CSD Crystal Packing Similarity* tool [85]. For this purpose, 15,5543 organic compounds were extracted from the CSD based on their composition and data quality. Then, a freely available algorythm published in 2010 [314] was used to cleave up to three acyclic single bonds between functional groups as demonstrated in Figure 14. Comparison of the *Crystal Packing Similarity* between molecular clusters formed by the "Value" molecular parts, allows us to reveal the most frequent terminal transformations of the "Key" groups and to evaluate their effect of the packing and isostructurality. For example, substitution of the methyl group to the chloro-, azide- or iodo- groups gives isostructural pairs in nearly 30% of cases.

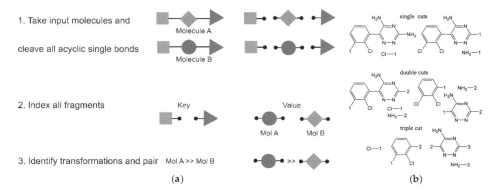

1. Take input molecules and cleave all acyclic single bonds

2. Index all fragments

3. Identify transformations and pair   Mol A >> Mol B

(a)

(b)

Figure 14. (**a**) Flowchart of the algorythm of bond cleavage reported at Refs. [85,314] and (**b**) some fragments formed from single, double, and triple cuts of lamotrigine.

Recently possibility of prediction of H-bonded motifs using a topological approach was demonstrated for some families of organic molecules with a large number of X-rayed representatives [315,316]. This method schematically represented in Figure 15 utilizes the idea that (i) only a limited number of architectures can be formed from a particular building block with a given connectivity, (ii) overall topology of a network depends on connectivity of building blocks, and (iii) propensities of various networks appearance are not equal. While relations between connectivity of an organic molecule and resulting network and corresponding propensities of their appearance have already been published [315,317], this approach does not allow evaluating the most abundant molecular connectivity of insufficiently studied families of organic compounds. Recently, maps of electrostatic potential, the *Full Interaction Maps*, and the *H-bonding Propensities* were compared as the tools to overcome this problem for heterocycle-1-carbohydrazoneamides [194]. Knowledge-based approaches demonstrated competition between various nitrogen atoms to act as acceptors of H-bonding with the donor -NH_2 group and allowed evaluating local molecular connectivity. Experimental crystal structures realize one of highly abundant theoretically predicted topologies of an H-bonded network.

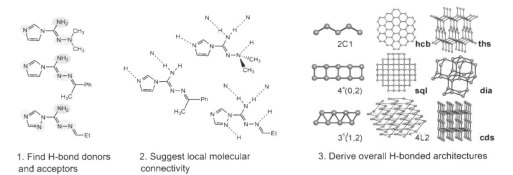

1. Find H-bond donors and acceptors

2. Suggest local molecular connectivity

3. Derive overall H-bonded architectures

Figure 15. Flowchart of the algorithm to estimate H-bonding architectures of organic molecules.

Tilbury, et al. suggested to combine the *H-bond propensity* based prediction with COSMO-RS theory [318,319] to predict drug substance hydrate formation [320]. Hydrate formation probability is estimated based on propensities calculated for each donor/acceptor groups in drug...drug or drug...water pairs. The most favorable hydrate formation has the maximal difference between these values. Although within this model relative strength of different functional groups to interact with water is displayed, this model does not take steric effects and intramolecular interactions into account. Additional quantum mechanics approach to eliminate donor and acceptor groups that are unable to form intermolecular interactions and to calculate the most favorable molecular and hydrate conformations allowed to improve predictive fidelity of calculations.

Chandra, et al. [225] applied CSD to enhance the solubility of telmisartan (TEL), a low soluble antihypertensive drug. First, the possibility of TEL to form likely intermolecular interactions with sulfobutylether beta-cyclodextrin was confirmed using H-bonding propensities tool. Then, the best docking pose of an inclusion complex of TEL in cyclodextrine was found with GOLD. The predicted binding mode was in line with the experimental spectra for the inclusion complex, which demonstrated enhanced solubility and dissolution rate of TEL.

Correlation between solubility and various parameters describing dimethyl sulfoxide (DMSO) role in crystals structures was carried out by Spitery, et al. [294]. They analyzed various parameters of hydrogen bonding of DMSO molecules as obtained from crystal structures reported in the CSD with solubility estimated using Chem3D [321], and found a negative correlation between the number of interactions the solvent is involved in and solubility of a solid.

8. Conclusions

The Cambridge Crystallographic Data Center holds a unique position of a curator of the Cambridge Crystallographic Database that now contains more than 10,000,00 crystal structures and a software developer for the analysis of collected data. This software allows the visualization of complex statistical data in a click-of-a-button manner (*Full Interaction Maps, Conformer Generator, Mogul analysis, Solvate Analyzer, BFDH predictions*), and investigation of local connectivity of functional groups and small molecules (*H-Bond Propensities, Co-Crystal Former, Motif Search*) that became possible due to the network between chemical diagrams and automatically calculating interatomic and intermolecular distances. Despite the complexity of the utilized algorithms, big data analysis, and their sophisticated processing, the software remains visible, user-friendly, and available for each CSD user. Thus, its' visibility and flexibility combined with exhaustive free documentation (see user guides, how-to-video, tutorials, and other materials at [322]) provide the basis for successful application in crystal engineering, material chemistry, and chemoinformatics.The diversity of CSD-supported studies described in this review indicate the great potential of the knowledge-based software for future development in these and associated fields, thus, the development of links with other molecular and crystallographic databases and repositories of properties would be advantageous.

The combination of the CSD tools with external software similarly with UNI and MOPAC calculations incorporated within the CSD-Enterprise seems to be very prospective for future development. DFT calculations, especially periodic ones, could provide further insight into the field of coordination polymers with numerous applications in gas storage and separation, catalysis, and spintronics. Implementation of graph theory to polytopic molecules and ligands could expand data about the local (molecular) connectivity to knowledge about possible and most abundant coordination and H-bonded architectures. The future of this field still possesses many challenges, but there is no denying that the value of the Cambridge Structural Database and associated software for the knowledge-based analysis in the filed of composition–structure–properties studies can not be overestimated.

Funding: This work was supported by the Ministry of Science and Higher Education of the Russian Federation.

Acknowledgments: The author is grateful to A. A. Korlyukov for fruitful discussion of the text.

Conflicts of Interest: The author declares no conflict of interest.

References

1. Allen, F.; Kennard, O.; Watson, D.; Brammer, L.; Orpen, A.; Taylor, R. Tables of Bond Lengths Determined by X-Ray and Neutron-Diffraction .1. Bond Lengths in Organic-Compounds. *J. Chem. Soc.-Perkin Trans. 2* **1987**, S1–S19. [CrossRef]

2. Orpen, A.; Brammer, L.; Allen, F.; Kennard, O.; Watson, D.; Taylor, R. Tables of Bond Lengths\Determined by X-Ray and Neutron-Diffraction .2. Organometallic Compounds and Co-Ordination Complexes of the D-Block and F-Block Metals. *J. Chem. Soc.-Dalton Trans.* **1989**, S1–S83. [CrossRef]

3. Allen, F.H.; Kennard, O.; Taylor, R. Systematic analysis of structural data as a research technique in organic chemistry. *Acc. Chem. Res.* **1983**, *16*, 146–153. [CrossRef]

4. Allen, F.H.; Kirby, A.J. Bond length and reactivity. Variable length of the carbon-oxygen single bond. *J. Am. Chem. Soc.* **1984**, *106*, 6197–6200. [CrossRef]

5. Kitaigorodskii, A.I. *Molecular Crystals and Molecules*; Academic Press: New York, NY, USA, 1973.

6. Desiraju, G.R.; Parshall, G.W. *Crystal Engineering: The Design of Organic Solids*; Elsevier: Amsterdam, The Netherlands, 1989.

7. Lehn, J.-M. *Supramolecular Chemistry: Concepts and Perspectives*; Wiley-VCH: Weinheim, Germany, 1995.

8. Nangia, A.K.; Desiraju, G.R. Crystal Engineering: An Outlook for the Future. *Angew. Chem. Int. Ed.* **2019**, *58*, 4100–4107. [CrossRef]

9. Corpinot, M.K.; Bučar, D.-K. A Practical Guide to the Design of Molecular Crystals. *Cryst. Growth Des.* **2019**, *19*, 1426–1453. [CrossRef]

10. Duggirala, N.K.; Perry, M.L.; Almarsson, Ö.; Zaworotko, M.J. Pharmaceutical cocrystals: Along the path to improved medicines. *Chem. Commun.* **2015**, *52*, 640–655. [CrossRef]

11. Zhang, J.; Xu, W.; Sheng, P.; Zhao, G.; Zhu, D. Organic Donor–Acceptor Complexes as Novel Organic Semiconductors. *Acc. Chem. Res.* **2017**, *50*, 1654–1662. [CrossRef]

12. Shi, P.-P.; Tang, Y.-Y.; Li, P.-F.; Liao, W.-Q.; Wang, Z.-X.; Ye, Q.; Xiong, R.-G. Symmetry breaking in molecular ferroelectrics. *Chem. Soc. Rev.* **2016**, *45*, 3811–3827. [CrossRef]

13. Turkoglu, G.; Cinar, M.E.; Ozturk, T. Triarylborane-Based Materials for OLED Applications. *Molecules* **2017**, *22*, 1522. [CrossRef]

14. Kidyarov, B.I. Comparative Interrelationship of the Structural, Nonlinear-Optical and Other Acentric Properties for Oxide, Borate and Carbonate Crystals. *Crystals* **2017**, *7*, 109. [CrossRef]

15. Quan, L.N.; Rand, B.P.; Friend, R.H.; Mhaisalkar, S.G.; Lee, T.-W.; Sargent, E.H. Perovskites for Next-Generation Optical Sources. *Chem. Rev.* **2019**, *119*, 7444–7477. [CrossRef] [PubMed]

16. Mackenzie, C.F.; Spackman, P.R.; Jayatilaka, D.; Spackman, M.A. CrystalExplorer model energies and energy frameworks: Extension to metal coordination compounds, organic salts, solvates and open-shell systems. *IUCrJ* **2017**, *4*, 575–587. [CrossRef] [PubMed]

17. Carugo, O.; Blatova, O.A.; Medrish, E.O.; Blatov, V.A.; Proserpio, D.M. Packing topology in crystals of proteins and small molecules: A comparison. *Sci. Rep.* **2017**, *7*, 13209. [CrossRef]

18. Taylor, R.; Wood, P.A. A Million Crystal Structures: The Whole Is Greater than the Sum of Its Parts. *Chem. Rev.* **2019**, *119*, 9247–9477. [CrossRef] [PubMed]

19. Groom, C.R.; Allen, F.H. The Cambridge Structural Database in Retrospect and Prospect. *Angew. Chem. Int. Ed.* **2014**, *53*, 662–671. [CrossRef]

20. Groom, C.R.; Bruno, I.J.; Lightfoot, M.P.; Ward, S.C. The Cambridge Structural Database. *Acta Cryst. Sect. B* **2016**, *72*, 171–179. [CrossRef] [PubMed]

21. Belsky, A.; Hellenbrandt, M.; Karen, V.L.; Luksch, P. New developments in the Inorganic Crystal Structure Database (ICSD): Accessibility in support of materials research and design. *Acta Cryst. Sect. B* **2002**, *58*, 364–369. [CrossRef] [PubMed]

22. Berman, H.M.; Westbrook, J.; Feng, Z.; Gilliland, G.; Bhat, T.N.; Weissig, H.; Shindyalov, I.N.; Bourne, P.E. The Protein Data Bank. *Nucleic Acids Res.* **2000**, *28*, 235–242. [CrossRef]

23. Peresypkina, E.V.; Blatov, V.A. Topology of molecular packings in organic crystals. *Acta Cryst. Sect. B* **2000**, *56*, 1035–1045. [CrossRef]

24. Peresypkina, E.V.; Blatov, V.A. Molecular coordination numbers in crystal structures of organic compounds. *Acta Cryst. Sect. B* **2000**, *56*, 501–511. [CrossRef] [PubMed]

25. Pathaneni, S.S.; Desiraju, G.R. Database analysis of Au··· Au interactions. *J. Chem. Soc. Dalton Trans.* **1993**, 319–322. [CrossRef]

26. Terrasson, V.; Planas, J.G.; Prim, D.; Teixidor, F.; Viñas, C.; Light, M.E.; Hursthouse, M.B. General Access to Aminobenzyl-o-carboranes as a New Class of Carborane Derivatives: Entry to Enantiopure Carborane–Amine Combinations. *Chem.—A Eur. J.* **2009**, *15*, 12030–12042. [CrossRef] [PubMed]

27. Thomas, S.P.; Pavan, M.S.; Row, T.N.G. Experimental evidence for 'carbon bonding' in the solid state from charge density analysis. *Chem. Commun.* **2013**, *50*, 49–51. [CrossRef] [PubMed]

28. Alkorta, I.; Del Bene, J.E.; Elguero, J. $H_2XP:OH_2$ Complexes: Hydrogen vs. Pnicogen Bonds. *Crystals* **2016**, *6*, 19. [CrossRef]

29. Lodochnikova, O.A.; Latypova, L.Z.; Madzhidov, T.I.; Chmutova, G.A.; Voronina, J.K.; Gubaidullin, A.T.; Kurbangalieva, A.R. "Lp··· synthon" interaction as a reason for the strong amplification of synthon-forming hydrogen bonds. *CrystEngComm* **2019**, *21*, 1499–1511. [CrossRef]

30. Taylor, R. Which intermolecular interactions have a significant influence on crystal packing? *CrystEngComm* **2014**, *16*, 6852–6865. [CrossRef]

31. Taylor, R. It Isn't, It Is: The C–H···X (X = O, N, F, Cl) Interaction Really Is Significant in Crystal Packing. *Cryst. Growth Des.* **2016**, *16*, 4165–4168. [CrossRef]

32. Jelsch, C.; Ejsmont, K.; Huder, L. The enrichment ratio of atomic contacts in crystals, an indicator derived from the Hirshfeld surface analysis. *IUCrJ* **2014**, *1*, 119–128. [CrossRef]

33. Jelsch, C.; Bibila Mayaya Bisseyou, Y. Atom interaction propensities of oxygenated chemical functions in crystal packings. *IUCrJ* **2017**, *4*, 158–174. [CrossRef]

34. Buergi, H.B.; Dunitz, J.D. From crystal statics to chemical dynamics. *Acc. Chem. Res.* **1983**, *16*, 153–161. [CrossRef]

35. Bürgi, H.-B.; Dunitz, J.D. *Structure Correlation*; Wiley-VCH: Weinheim, Germany, 1994.

36. Vologzhanina, A.V.; Korlyukov, A.A.; Antipin, M.Y. Special features of intermolecular bonding A···D (A = Si, Ge and D = nucleophile) in crystal structures. *Acta Cryst. Sect. B* **2008**, *64*, 448–455. [CrossRef] [PubMed]

37. Bruno, I.J.; Cole, J.C.; Kessler, M.; Luo, J.; Motherwell, W.D.S.; Purkis, L.H.; Smith, B.R.; Taylor, R.; Cooper, R.I.; Harris, S.E.; et al. Retrieval of Crystallographically-Derived Molecular Geometry Information. *J. Chem. Inf. Comput. Sci.* **2004**, *44*, 2133–2144. [CrossRef] [PubMed]

38. Bruno, I.J.; Cole, J.C.; Lommerse, J.P.M.; Rowland, R.S.; Taylor, R.; Verdonk, M.L. IsoStar: A library of information about nonbonded interactions. *J. Comput. Aided Mol. Des.* **1997**, *11*, 525–537. [CrossRef] [PubMed]

39. McKenzie, J.; Feeder, N.; Hunter, C.A. H-bond competition experiments in solution and the solid state. *CrystEngComm* **2016**, *18*, 394–397. [CrossRef]

40. Mascal, M.; Infantes, L.; Chisholm, J. Water Oligomers in Crystal Hydrates—What's News and What Isn't? *Angew. Chem. Int. Ed.* **2006**, *45*, 32–36. [CrossRef] [PubMed]

41. Manna, U.; Halder, S.; Das, G. Ice-like Cyclic Water Hexamer Trapped within a Halide Encapsulated Hexameric Neutral Receptor Core: First Crystallographic Evidence of a Water Cluster Confined within a Receptor-Anion Capsular Assembly. *Cryst. Growth Des.* **2018**, *18*, 1818–1825. [CrossRef]

42. Desiraju, G.R. Supramolecular Synthons in Crystal Engineering—A New Organic Synthesis. *Angew. Chem. Int. Ed.* **1995**, *34*, 2311–2327. [CrossRef]

43. Aakeröy, C.B. Crystal Engineering: Strategies and Architectures. *Acta Cryst. Sect. B* **1997**, *53*, 569–586. [CrossRef]

44. Hollingsworth, M.D. Crystal Engineering: From Structure to Function. *Science* **2002**, *295*, 2410–2413.

45. Nagarathinam, M.; Peedikakkal, A.M.P.; Vittal, J.J. Stacking of double bonds for photochemical [2+2] cycloaddition reactions in the solid state. *Chem. Commun.* **2008**, 5277–5288. [CrossRef] [PubMed]

46. Ramamurthy, V.; Sivaguru, J. Supramolecular Photochemistry as a Potential Synthetic Tool: Photocycloaddition. *Chem. Rev.* **2016**, *116*, 9914–9993. [CrossRef] [PubMed]

47. Kreuer, K.-D. Proton Conductivity: Materials and Applications. *Chem. Mater.* **1996**, *8*, 610–641. [CrossRef]

48. Horiuchi, S.; Kumai, R.; Tokura, Y. Hydrogen-Bonding Molecular Chains for High-Temperature Ferroelectricity. *Adv. Mater.* **2011**, *23*, 2098–2103. [CrossRef] [PubMed]

49. Horiuchi, S.; Noda, Y.; Hasegawa, T.; Kagawa, F.; Ishibashi, S. Correlated Proton Transfer and Ferroelectricity along Alternating Zwitterionic and Nonzwitterionic Anthranilic Acid Molecules. *Chem. Mater.* **2015**, *27*, 6193–6197. [CrossRef]

50. Horiuchi, S.; Kagawa, F.; Hatahara, K.; Kobayashi, K.; Kumai, R.; Murakami, Y.; Tokura, Y. Above-room-temperature ferroelectricity and antiferroelectricity in benzimidazoles. *Nat. Commun.* **2012**, *3*, 1308. [CrossRef] [PubMed]

51. Owczarek, M.; Hujsak, K.A.; Ferris, D.P.; Prokofjevs, A.; Majerz, I.; Szklarz, P.; Zhang, H.; Sarjeant, A.A.; Stern, C.L.; Jakubas, R.; et al. Flexible ferroelectric organic crystals. *Nat. Commun.* **2016**, *7*, 13108. [CrossRef] [PubMed]

52. Borel, C.; Larsson, K.; Håkansson, M.; Olsson, B.E.; Bond, A.D.; Öhrström, L. Oxalate- and Squarate-Biimidazole Supramolecular Synthons: Hydrogen-Bonded Networks Based on [Co(H$_2$biimidazole)$_3$]$^{3+}$. *Cryst. Growth Des.* **2009**, *9*, 2821–2827. [CrossRef]

53. Pallipurath, A.R.; Civati, F.; Eziashi, M.; Omar, E.; McArdle, P.; Erxleben, A. Tailoring Cocrystal and Salt Formation and Controlling the Crystal Habit of Diflunisal. *Cryst. Growth Des.* **2016**, *16*, 6468–6478. [CrossRef]

54. George, F.; Norberg, B.; Wouters, J.; Leyssens, T. Structural Investigation of Substituent Effect on Hydrogen Bonding in (S)-Phenylglycine Amide Benzaldimines. *Cryst. Growth Des.* **2015**, *15*, 4005–4019. [CrossRef]

55. Bernstein, J. *Polymorphism in Molecular Crystals*; Oxford University Press: Oxford, UK, 2007.

56. Bauer, J.; Spanton, S.; Henry, R.; Quick, J.; Dziki, W.; Porter, W.; Morris, J. Ritonavir: An Extraordinary Example of Conformational Polymorphism. *Pharm. Res.* **2001**, *18*, 859–866. [CrossRef] [PubMed]

57. Dalinger, I.L.; Vatsadze, I.A.; Shkineva, T.K.; Kormanov, A.V.; Struchkova, M.I.; Suponitsky, K.Yu.; Bragin, A.A.; Monogarov, K.A.; Sinditskii, V.P.; Sheremetev, A.B. Novel Highly Energetic Pyrazoles: N-Trinitromethyl-Substituted Nitropyrazoles. *Chem.—Asian J.* **2015**, *10*, 1987–1996. [CrossRef] [PubMed]

58. Dalinger, I.L.; Kormanov, A.V.; Suponitsky, K.Y.; Muravyev, N.V.; Sheremetev, A.B. Pyrazole-Tetrazole Hybrid with Trinitromethyl, Fluorodinitromethyl, or (Difluoroamino)dinitromethyl Groups: High-Performance Energetic Materials. *Chem.—Asian J.* **2018**, *13*, 1165–1172. [CrossRef] [PubMed]

59. Dalinger, I.L.; Serushkina, O.V.; Muravyev, N.V.; Meerov, D.B.; Miroshnichenko, E.A.; Kon'kova, T.S.; Suponitsky, K.Y.; Vener, M.V.; Sheremetev, A.B. Azasydnone—Novel "green" building block for designing high energetic compounds. *J. Mater. Chem. A* **2018**, *6*, 18669–18676. [CrossRef]

60. Bredikhin, A.A.; Bredikhina, Z.A.; Zakharychev, D.V. Crystallization of chiral compounds: Thermodynamical, structural and practical aspects. *Mendeleev Commun.* **2012**, *22*, 171–180. [CrossRef]

61. Bredikhin, A.A.; Zakharychev, D.V.; Bredikhina, Z.A.; Kurenkov, A.V.; Krivolapov, D.B.; Gubaidullin, A.T. Spontaneous Resolution of Chiral 3-(2,3-Dimethylphenoxy)propane-1,2-diol under the Circumstances of an Unusual Diversity of Racemic Crystalline Modifications. *Cryst. Growth Des.* **2017**, *17*, 4196–4206. [CrossRef]

62. Kinbara, K.; Hashimoto, Y.; Sukegawa, M.; Nohira, H.; Saigo, K. Crystal Structures of the Salts of Chiral Primary Amines with Achiral Carboxylic Acids: Recognition of the Commonly-Occurring Supramolecular Assemblies of Hydrogen-Bond Networks and Their Role in the Formation of Conglomerates. *J. Am. Chem. Soc.* **1996**, *118*, 3441–3449. [CrossRef]

63. Haynes, D.A.; Chisholm, J.A.; Jones, W.; Motherwell, W.D.S. Supramolecular synthon competition in organic sulfonates: A CSD survey. *CrystEngComm* **2004**, *6*, 584–588. [CrossRef]

64. da Silva, C.C.P.; de Oliveira, R.; Tenorio, J.C.; Honorato, S.B.; Ayala, A.P.; Ellena, J. The Continuum in 5-Fluorocytosine. Toward Salt Formation. *Cryst. Growth Des.* **2013**, *13*, 4315–4322. [CrossRef]

65. Chisholm, J.A.; Motherwell, S. COMPACK: A program for identifying crystal structure similarity using distances. *J. Appl. Cryst.* **2005**, *38*, 228–231. [CrossRef]

66. Wanat, M.; Malinska, M.; Kutner, A.; Wozniak, K. Effect of Vitamin D Conformation on Interactions and Packing in the Crystal Lattice. *Cryst. Growth Des.* **2018**, *18*, 3385–3396. [CrossRef]

67. Infantes, L.; Motherwell, S. Water clusters in organic molecular crystals. *CrystEngComm* **2002**, *4*, 454–461. [CrossRef]

68. Blatov, V.A.; O'Keeffe, M.; Proserpio, D.M. Vertex-, face-, point-, Schläfli-, and Delaney-symbols in nets, polyhedra and tilings: Recommended terminology. *CrystEngComm* **2009**, *12*, 44–48. [CrossRef]

69. Batten, S.R.; Champness, N.R.; Chen, X.-M.; Garcia-Martinez, J.; Kitagawa, S.; Ohrstrom, L.; O'Keeffe, M.; Suh, M.P.; Reedijk, J. Terminology of metal–organic frameworks and coordination polymers (IUPAC Recommendations 2013). *Pure Appl. Chem.* **2013**, *85*, 1715–1724. [CrossRef]

70. Ohrstrom, L. Designing, Describing and Disseminating New Materials by using the Network Topology Approach. *Chem.—Eur. J.* **2016**, *22*, 13758–13763. [CrossRef] [PubMed]

71. Bonneau, C.; O'Keeffe, M.; Proserpio, D.M.; Blatov, V.A.; Batten, S.R.; Bourne, S.A.; Lah, M.S.; Eon, J.-G.; Hyde, S.T.; Wiggin, S.B.; et al. Deconstruction of Crystalline Networks into Underlying Nets: Relevance for Terminology Guidelines and Crystallographic Databases. *Cryst. Growth Des.* **2018**, *18*, 3411–3418. [CrossRef]

72. Haneline, M.R.; Tsunoda, M.; Gabbaï, F.P. π-Complexation of Biphenyl, Naphthalene, and Triphenylene to Trimeric Perfluoro-ortho-phenylene Mercury. Formation of Extended Binary Stacks with Unusual Luminescent Properties. *J. Am. Chem. Soc.* **2002**, *124*, 3737–3742. [CrossRef] [PubMed]

73. Burress, C.; Elbjeirami, O.; Omary, M.A.; Gabbaï, F.P. Five-Order-of-Magnitude Reduction of the Triplet Lifetimes of N-Heterocycles by Complexation to a Trinuclear Mercury Complex. *J. Am. Chem. Soc.* **2005**, *127*, 12166–12167. [CrossRef] [PubMed]

74. Taylor, T.J.; Burress, C.N.; Gabbaï, F.P. Lewis Acidic Behavior of Fluorinated Organomercurials. *Organometallics* **2007**, *26*, 5252–5263. [CrossRef]

75. Ferraris, J.; Cowan, D.O.; Walatka, V.; Perlstein, J.H. Electron transfer in a new highly conducting donor-acceptor complex. *J. Am. Chem. Soc.* **1973**, *95*, 948–949. [CrossRef]

76. Giri, G.; Verploegen, E.; Mannsfeld, S.C.B.; Atahan-Evrenk, S.; Kim, D.H.; Lee, S.Y.; Becerril, H.A.; Aspuru-Guzik, A.; Toney, M.F.; Bao, Z. Tuning charge transport in solution-sheared organic semiconductors using lattice strain. *Nature* **2011**, *480*, 504–508. [CrossRef] [PubMed]

77. Varghese, S.; Das, S. Role of Molecular Packing in Determining Solid-State Optical Properties of π-Conjugated Materials. *J. Phys. Chem. Lett.* **2011**, *2*, 863–873. [CrossRef] [PubMed]

78. Zou, T.; Wang, X.; Ju, H.; Zhao, L.; Guo, T.; Wu, W.; Wang, H. Controllable Molecular Packing Motif and Overlap Type in Organic Nanomaterials for Advanced Optical Properties. *Crystals* **2018**, *8*, 22. [CrossRef]

79. Larin, A.A.; Muravyev, N.V.; Pivkina, A.N.; Suponitsky, K.Y.; Ananyev, I.V.; Khakimov, D.V.; Fershtat, L.L.; Makhova, N.N. Assembly of Tetrazolylfuroxan Organic Salts: Multipurpose Green Energetic Materials with High Enthalpies of Formation and Excellent Detonation Performance. *Chem.—Eur. J.* **2019**, *25*, 4225–4233. [CrossRef] [PubMed]

80. Dolgushin, F.M.; Smol'yakov, A.F.; Suponitsky, K.Y.; Vologzhanina, A.V.; Fedyanin, I.V.; Shishkina, S.V. Intermolecular interactions in polymorphs of the cyclic trimeric perfluoro-ortho-phenylene mercury from geometric, energetic and AIM viewpoints: DFT study and Hirshfeld surface analysis. *Struct. Chem.* **2016**, *27*, 37–49. [CrossRef]

81. Torubaev, Y.V.; Rai, D.K.; Skabitsky, I.V.; Pakhira, S.; Dmitrienko, A. Energy framework approach to the supramolecular reactions: Interplay of the secondary bonding interaction in Ph_2E_2 (E = Se, Te)/p-I-C_6F_4-I co-crystals. *New J. Chem.* **2019**, *43*, 7941–7949. [CrossRef]

82. Martins, F.T.; Paparidis, N.; Doriguetto, A.C.; Ellena, J. Crystal Engineering of an Anti-HIV Drug Based on the Recognition of Assembling Molecular Frameworks. *Cryst. Growth Des.* **2009**, *9*, 5283–5292. [CrossRef]

83. Pogoda, D.; Janczak, J.; Videnova-Adrabinska, V. New polymorphs of an old drug: Conformational and synthon polymorphism of 5-nitrofurazone. *Acta Cryst. Sect. B* **2016**, *72*, 263–273. [CrossRef]

84. Reilly, A.M.; Cooper, R.I.; Adjiman, C.S.; Bhattacharya, S.; Boese, A.D.; Brandenburg, J.G.; Bygrave, P.J.; Bylsma, R.; Campbell, J.E.; Car, R.; et al. Report on the sixth blind test of organic crystal structure prediction methods. *Acta Cryst. Sect. B* **2016**, *72*, 439–459. [CrossRef]

85. Giangreco, I.; Cole, J.C.; Thomas, E. Mining the Cambridge Structural Database for Matched Molecular Crystal Structures: A Systematic Exploration of Isostructurality. *Cryst. Growth Des.* **2017**, *17*, 3192–3203. [CrossRef]

86. Carletta, A.; Colaço, M.; Mouchet, S.R.; Plas, A.; Tumanov, N.; Fusaro, L.; Champagne, B.; Lanners, S.; Wouters, J. Tetraphenylborate Anion Induces Photochromism in N-Salicylideneamino-1-alkylpyridinium Derivatives Through Formation of Tetra-Aryl Boxes. *J. Phys. Chem. C* **2018**, *122*, 10999–11007. [CrossRef]

87. Nath, N.K.; Saha, B.K.; Nangia, A. Isostructural polymorphs of triiodophloroglucinol and triiodoresorcinol. *New J. Chem.* **2008**, *32*, 1693–1701. [CrossRef]

88. Moorthy, J.N.; Mandal, S.; Venugopalan, P. Hydrogen-Bonded Helical Self-Assembly of Sterically-Hindered Benzyl Alcohols: Rare Isostructurality and Synthon Equivalence Between Alcohols and Acids. *Cryst. Growth Des.* **2012**, *12*, 2942–2947. [CrossRef]

89. Reddy, C.M.; Kirchner, M.T.; Gundakaram, R.C.; Padmanabhan, K.A.; Desiraju, G.R. Isostructurality, Polymorphism and Mechanical Properties of Some Hexahalogenated Benzenes: The Nature of Halogen···Halogen Interactions. *Chem.—Eur. J.* **2006**, *12*, 2222–2234. [CrossRef] [PubMed]

90. Asmadi, A.; Kendrick, J.; Leusen, F.J.J. Crystal Structure Prediction and Isostructurality of Three Small Molecules. *Chem.—Eur. J.* **2010**, *16*, 12701–12709. [CrossRef]

91. Owczarzak, A.M.; Kourkoumelis, N.; Hadjikakou, S.K.; Kubicki, M. The impact of the anion size on the crystal packing in 2-mercaptopyrimidine halides; isostructurality and polymorphism. *CrystEngComm* **2013**, *15*, 3607–3614. [CrossRef]

92. Ravat, P.; SeethaLekshmi, S.; Biswas, S.N.; Nandy, P.; Varughese, S. Equivalence of Ethylene and Azo-Bridges in the Modular Design of Molecular Complexes: Role of Weak Interactions. *Cryst. Growth Des.* **2015**, *15*, 2389–2401. [CrossRef]

93. Nath, N.K.; Nangia, A. Isomorphous Crystals by Chloro–Methyl Exchange in Polymorphic Fuchsones. *Cryst. Growth Des.* **2012**, *12*, 5411–5425. [CrossRef]

94. Gonnade, R.G.; Bhadbhade, M.M.; Shashidhar, M.S. Crystal-to-crystal thermal phase transition amongst dimorphs of hexa-O-p-toluoyl-myo-inositol conserving two-dimensional isostructurality. *CrystEngComm* **2010**, *12*, 478–484. [CrossRef]

95. Gelbrich, T.; Hughes, D.S.; Hursthouse, M.B.; Threlfall, T.L. Packing similarity in polymorphs of sulfathiazole. *CrystEngComm* **2008**, *10*, 1328–1334. [CrossRef]

96. Panini, P.; Mohan, T.P.; Gangwar, U.; Sankolli, R.; Chopra, D. Quantitative crystal structure analysis of 1,3,4-thiadiazole derivatives. *CrystEngComm* **2013**, *15*, 4549–4564. [CrossRef]

97. Chopra, D.; Row, T.N.G. Evaluation of the interchangeability of C–H and C–F groups: Insights from crystal packing in a series of isomeric fluorinated benzanilides. *CrystEngComm* **2007**, *10*, 54–67. [CrossRef]

98. Thakuria, R.; Nath, N.K.; Roy, S.; Nangia, A. Polymorphism and isostructurality in sulfonylhydrazones. *CrystEngComm* **2014**, *16*, 4681–4690. [CrossRef]

99. Gelbrich, T.; Hursthouse, M.B. A versatile procedure for the identification, description and quantification of structural similarity in molecular crystals. *CrystEngComm* **2005**, *7*, 324–336. [CrossRef]

100. Dzyabchenko, A.V. Method of crystal-structure similarity searching. *Acta Cryst. Sect. B* **1994**, *50*, 414–425.

101. Suponitsky, K.Y.; Lyssenko, K.A.; Ananyev, I.V.; Kozeev, A.M.; Sheremetev, A.B. Role of Weak Intermolecular Interactions in the Crystal Structure of Tetrakis-furazano[3,4-c:3′,4′-g:3″,4″-k:3‴,4‴-o][1,2,5,6,9,10,13,14]octaazacyclohexadecine and Its Solvates. *Cryst. Growth Des.* **2014**, *14*, 4439–4449. [CrossRef]

102. Childs, S.L.; Wood, P.A.; Rodríguez-Hornedo, N.; Reddy, L.S.; Hardcastle, K.I. Analysis of 50 Crystal Structures Containing Carbamazepine Using the Materials Module of Mercury CSD. *Cryst. Growth Des.* **2009**, *9*, 1869–1888. [CrossRef]

103. Gelbrich, T.; Hursthouse, M.B. Systematic investigation of the relationships between 25 crystal structures containing the carbamazepine molecule or a close analogue: A case study of the XPac method. *CrystEngComm* **2006**, *8*, 448–460. [CrossRef]

104. Prohens, R.; Font-Bardia, M.; Barbas, R. Water wires in the nanoporous form II of carbamazepine: A single-crystal X-ray diffraction analysis. *CrystEngComm* **2013**, *15*, 845–847. [CrossRef]

105. Zhong, Z.; Yang, X.; Wang, B.-H.; Yao, Y.-F.; Guo, B.; Yu, L.; Huang, Y.; Xu, J. Solvent-polymer guest exchange in a carbamazepine inclusion complex: Structure, kinetics and implication for guest selection. *CrystEngComm* **2019**, *21*, 2164–2173. [CrossRef]

106. Kennedy, A.R.; Morrison, C.A.; Briggs, N.E.B.; Arbuckle, W. Density and Stability Differences Between Enantiopure and Racemic Salts: Construction and Structural Analysis of a Systematic Series of Crystalline Salt Forms of Methylephedrine. *Cryst. Growth Des.* **2011**, *11*, 1821–1834. [CrossRef]

107. Tumanov, N.A.; Myz, S.A.; Shakhtshneider, T.P.; Boldyreva, E.V. Are meloxicam dimers really the structure-forming units in the 'meloxicam–carboxylic acid' co-crystals family? Relation between crystal structures and dissolution behaviour. *CrystEngComm* **2012**, *14*, 305–313. [CrossRef]

108. Ueto, T.; Takata, N.; Muroyama, N.; Nedu, A.; Sasaki, A.; Tanida, S.; Terada, K. Polymorphs and a Hydrate of Furosemide–Nicotinamide 1:1 Cocrystal. *Cryst. Growth Des.* **2012**, *12*, 485–494. [CrossRef]

109. Galcera, J.; Friščić, T.; Molins, E.; Jones, W. Isostructurality in three-component crystals achieved by the combination of persistent hydrogen bonding motifs and solvent inclusion. *CrystEngComm* **2013**, *15*, 1332–1338. [CrossRef]

110. Briggs, N.E.; Kennedy, A.R.; Morrison, C.A. 42 salt forms of tyramine: Structural comparison and the occurrence of hydrate formation. *Acta Cryst. Sect. B* **2012**, *68*, 453–464. [CrossRef] [PubMed]

111. Salbego, P.R.S.; Bender, C.R.; Hörner, M.; Zanatta, N.; Frizzo, C.P.; Bonacorso, H.G.; Martins, M.A.P. Insights on the Similarity of Supramolecular Structures in Organic Crystals Using Quantitative Indexes. *ACS Omega* **2018**, *3*, 2569–2578. [CrossRef] [PubMed]

112. Surov, A.O.; Simagina, A.A.; Manin, N.G.; Kuzmina, L.G.; Churakov, A.V.; Perlovich, G.L. Fenamate Cocrystals with 4,4'-Bipyridine: Structural and Thermodynamic Aspects. *Cryst. Growth Des.* **2015**, *15*, 228–238. [CrossRef]

113. Sládková, V.; Skalická, T.; Skořepová, E.; Čejka, J.; Eigner, V.; Kratochvíl, B. Systematic solvate screening of trospium chloride: Discovering hydrates of a long-established pharmaceutical. *CrystEngComm* **2015**, *17*, 4712–4721. [CrossRef]

114. Chennuru, R.; Muthudoss, P.; Voguri, R.S.; Ramakrishnan, S.; Vishweshwar, P.; Babu, R.R.C.; Mahapatra, S. Iso-Structurality Induced Solid Phase Transformations: A Case Study with Lenalidomide. *Cryst. Growth Des.* **2017**, *17*, 612–628. [CrossRef]

115. Allen, F.H.; Groom, C.R.; Liebeschuetz, J.W.; Bardwell, D.A.; Olsson, T.S.G.; Wood, P.A. The Hydrogen Bond Environments of 1H-Tetrazole and Tetrazolate Rings: The Structural Basis for Tetrazole–Carboxylic Acid Bioisosterism. *J. Chem. Inf. Model.* **2012**, *52*, 857–866. [CrossRef]

116. Chernyshov, I.Y.; Vener, M.V.; Prikhodchenko, P.V.; Medvedev, A.G.; Lev, O.; Churakov, A.V. Peroxosolvates: Formation Criteria, H2O2 Hydrogen Bonding, and Isomorphism with the Corresponding Hydrates. *Cryst. Growth Des.* **2017**, *17*, 214–220. [CrossRef]

117. Cummings, M.D.; Sekharan, S. Structure-Based Macrocycle Design in Small-Molecule Drug Discovery and Simple Metrics to Identify Opportunities for Macrocyclization of Small-Molecule Ligands. *J. Med. Chem.* **2019**, *62*, 6843–6853. [CrossRef] [PubMed]

118. Gavezzotti, A. Are Crystal Structures Predictable? *Acc. Chem. Res.* **1994**, *27*, 309–314. [CrossRef]

119. Gavezzotti, A. The Crystal Packing of Organic Molecules: Challenge and Fascination Below 1000 Da. *Crystallogr. Rev.* **1998**, *7*, 5–121. [CrossRef]

120. Chemla, D.S.; Zyss, J. *Nonlinear Optical Properties of Organic Molecules and Crystals*; Elsevier: Amsterdam, The Netherlands, 1987.

121. Ostroverkhova, O. Organic Optoelectronic Materials: Mechanisms and Applications. *Chem. Rev.* **2016**, *116*, 13279–13412. [CrossRef] [PubMed]

122. Tyagi, N.; Sinha, N.; Yadav, H.; Kumar, B. Growth, morphology, structure and characterization of l-histidinium dihydrogen arsenate orthoarsenic acid single crystal. *Acta Cryst. Sect. B* **2016**, *72*, 593–601. [CrossRef] [PubMed]

123. Gao, Z.; Tian, X.; Zhang, J.; Wu, Q.; Lu, Q.; Tao, X. Large-Sized Crystal Growth and Electric-Elastic Properties of α-BaTeMo2O9 Single Crystal. *Cryst. Growth Des.* **2015**, *15*, 759–763. [CrossRef]

124. Chen, F.; Jiang, C.; Tian, S.; Yu, F.; Cheng, X.; Duan, X.; Wang, Z.; Zhao, X. Electroelastic Features of Piezoelectric Bi2ZnB2O7 Crystal. *Cryst. Growth Des.* **2018**, *18*, 3988–3996. [CrossRef]

125. Leclaire, N.A.; Li, M.; Véron, A.C.; Neels, A.; Heier, J.; Reimers, J.R.; Nüesch, F.A. Cyanine platelet single crystals: Growth, crystal structure and optical spectra. *Phys. Chem. Chem. Phys.* **2018**, *20*, 29166–29173. [CrossRef]

126. Ozdemir, R.; Park, S.; Deneme, İ.; Park, Y.; Zorlu, Y.; Ardic Alidagi, H.; Harmandar, K.; Kim, C.; Usta, H. Triisopropylsilylethynyl-substituted indenofluorenes: Carbonyl versus dicyanovinylene functionalization in one-dimensional molecular crystals and solution-processed n-channel OFETs. *Org. Chem. Front.* **2018**, *5*, 2912–2924. [CrossRef]

127. Glöcklhofer, F.; Morawietz, A.J.; Stöger, B.; Unterlass, M.M.; Fröhlich, J. Extending the Scope of a New Cyanation: Design and Synthesis of an Anthracene Derivative with an Exceptionally Low LUMO Level and Improved Solubility. *ACS Omega* **2017**, *2*, 1594–1600. [CrossRef] [PubMed]

128. Wang, C.; Dong, H.; Jiang, L.; Hu, W. Organic semiconductor crystals. *Chem. Soc. Rev.* **2018**, *47*, 422–500. [CrossRef] [PubMed]

129. Stevens, J.S.; Walczak, M.; Jaye, C.; Fischer, D.A. In Situ Solid-State Reactions Monitored by X-ray Absorption Spectroscopy: Temperature-Induced Proton Transfer Leads to Chemical Shifts. *Chem.—Eur. J.* **2016**, *22*, 15600–15604. [CrossRef] [PubMed]

130. Liu, G.; Liu, J.; Liu, Y.; Tao, X. Oriented Single-Crystal-to-Single-Crystal Phase Transition with Dramatic Changes in the Dimensions of Crystals. *J. Am. Chem. Soc.* **2014**, *136*, 590–593. [CrossRef] [PubMed]

131. Chou, C.-M.; Nobusue, S.; Saito, S.; Inoue, D.; Hashizume, D.; Yamaguchi, S. Highly bent crystals formed by restrained π-stacked columns connected via alkylene linkers with variable conformations. *Chem. Sci.* **2015**, *6*, 2354–2359. [CrossRef] [PubMed]

132. Commins, P.; Desta, I.T.; Karothu, D.P.; Panda, M.K.; Naumov, P. Crystals on the move: Mechanical effects in dynamic solids. *Chem. Commun.* **2016**, *52*, 13941–13954. [CrossRef] [PubMed]

133. Boldyreva, E.V. High-pressure diffraction studies of molecular organic solids. A personal view. *Acta Cryst. Sect. A* **2008**, *64*, 218–231. [CrossRef]

134. Bull, C.L.; Flowitt-Hill, G.; de Gironcoli, S.; Küçükbenli, E.; Parsons, S.; Pham, C.H.; Playford, H.Y.; Tucker, M.G. ζ-Glycine: Insight into the mechanism of a polymorphic phase transition. *IUCrJ* **2017**, *4*, 569–574. [CrossRef]

135. Giordano, N.; Beavers, C.M.; Kamenev, K.V.; Marshall, W.G.; Moggach, S.A.; Patterson, S.D.; Teat, S.J.; Warren, J.E.; Wood, P.A.; Parsons, S. High-pressure polymorphism in L-threonine between ambient pressure and 22 GPa. *CrystEngComm* **2019**, *21*, 4444–4456. [CrossRef]

136. Sun, C.; Grant, D.J.W. Influence of Crystal Shape on the Tableting Performance of L-Lysine Monohydrochloride Dihydrate. *J. Pharm. Sci.* **2001**, *90*, 569–579. [CrossRef]

137. Rosbottom, I.; Ma, C.Y.; Turner, T.D.; O'Connell, R.A.; Loughrey, J.; Sadiq, G.; Davey, R.J.; Roberts, K.J. Influence of Solvent Composition on the Crystal Morphology and Structure of p-Aminobenzoic Acid Crystallized from Mixed Ethanol and Nitromethane Solutions. *Cryst. Growth Des.* **2017**, *17*, 4151–4161. [CrossRef]

138. Zhao, P.; Liu, X.; Wang, L.; Gao, Z.; Yang, Y.; Hao, H.; Xie, C.; Bao, Y. Predicting the crystal habit of photoinitiator XBPO and elucidating the solvent effect on crystal faces. *CrystEngComm* **2019**, *21*, 2422–2430. [CrossRef]

139. O'Mahony, M.; Seaton, C.C.; Croker, D.M.; Veesler, S.; Rasmuson, Å.C.; Hodnett, B.K. Investigating the dissolution of the metastable triclinic polymorph of carbamazepine using in situ microscopy. *CrystEngComm* **2014**, *16*, 4133–4141. [CrossRef]

140. Vologzhanina, A.V.; Belov, A.S.; Novikov, V.V.; Dolganov, A.V.; Romanenko, G.V.; Ovcharenko, V.I.; Korlyukov, A.A.; Buzin, M.I.; Voloshin, Y.Z. Synthesis and Temperature-Induced Structural Phase and Spin Transitions in Hexadecylboron-Capped Cobalt(II) Hexachloroclathrochelate and Its Diamagnetic Iron(II)-Encapsulating Analogue. *Inorg. Chem.* **2015**, *54*, 5827–5838. [CrossRef] [PubMed]

141. Pavlov, A.A.; Nelyubina, Y.V.; Kats, S.V.; Penkova, L.V.; Efimov, N.N.; Dmitrienko, A.O.; Vologzhanina, A.V.; Belov, A.S.; Voloshin, Y.Z.; Novikov, V.V. Polymorphism in a Cobalt-Based Single-Ion Magnet Tuning Its Barrier to Magnetization Relaxation. *J. Phys. Chem. Lett.* **2016**, *7*, 4111–4116. [CrossRef] [PubMed]

142. Nguyen, T.T.H.; Rosbottom, I.; Marziano, I.; Hammond, R.B.; Roberts, K.J. Crystal Morphology and Interfacial Stability of RS-Ibuprofen in Relation to Its Molecular and Synthonic Structure. *Cryst. Growth Des.* **2017**, *17*, 3088–3099. [CrossRef]

143. Rosbottom, I.; Pickering, J.H.; Etbon, B.; Hammond, R.B.; Roberts, K.J. Examination of inequivalent wetting on the crystal habit surfaces of RS-ibuprofen using grid-based molecular modelling. *Phys. Chem. Chem. Phys.* **2018**, *20*, 11622–11633. [CrossRef]

144. Turner, T.D.; Hatcher, L.E.; Wilson, C.C.; Roberts, K.J. Habit Modification of the Active Pharmaceutical Ingredient Lovastatin Through a Predictive Solvent Selection Approach. *J. Pharm. Sci.* **2019**, *108*, 1779–1787. [CrossRef]

145. Gajda, R.; Domański, M.A.; Malinska, M.; Makal, A. Crystal morphology fixed by interplay of π-stacking and hydrogen bonds – the case of 1-hydroxypyrene. *CrystEngComm* **2019**, *21*, 1701–1717. [CrossRef]

146. Han, D.; Karmakar, T.; Bjelobrk, Z.; Gong, J.; Parrinello, M. Solvent-mediated morphology selection of the active pharmaceutical ingredient isoniazid: Experimental and simulation studies. *Chem. Eng. Sci.* **2019**, *204*, 320–328. [CrossRef]

147. Destri, G.L.; Marrazzo, A.; Rescifina, A.; Punzo, F. Crystal Morphologies and Polymorphs in Tolbutamide Microcrystalline Powder. *J. Pharm. Sci.* **2013**, *102*, 73–83. [CrossRef]

148. Thirupugalmani, K.; Venkatesh, M.; Karthick, S.; Maurya, K.K.; Vijayan, N.; Chaudhary, A.K.; Brahadeeswaran, S. Influence of polar solvents on growth of potentially NLO active organic single crystals of N-benzyl-2-methyl-4-nitroaniline and their efficiency in terahertz generation. *CrystEngComm* **2017**, *19*, 2623–2631. [CrossRef]

149. Croker, D.M.; Kelly, D.M.; Horgan, D.E.; Hodnett, B.K.; Lawrence, S.E.; Moynihan, H.A.; Rasmuson, Å.C. Demonstrating the Influence of Solvent Choice and Crystallization Conditions on Phenacetin Crystal Habit and Particle Size Distribution. *Org. Process Res. Dev.* **2015**, *19*, 1826–1836. [CrossRef]

150. Poornachary, S.K.; Lau, G.; Chow, P.S.; Tan, R.B.H.; George, N. The Effect and Counter-Effect of Impurities on Crystallization of an Agrochemical Active Ingredient: Stereochemical Rationalization and Nanoscale Crystal Growth Visualization. *Cryst. Growth Des.* **2011**, *11*, 492–500. [CrossRef]

151. Anuar, N.; Wan Daud, W.R.; Roberts, K.J.; Kamarudin, S.K.; Tasirin, S.M. Morphology and Associated Surface Chemistry of l-Isoleucine Crystals Modeled under the Influence of l-Leucine Additive Molecules. *Cryst. Growth Des.* **2012**, *12*, 2195–2203. [CrossRef]

152. Tang, Y.; Gao, J. Investigation of the Effects of Sodium Dicarboxylates on the Crystal Habit of Calcium Sulfate α-Hemihydrate. *Langmuir* **2017**, *33*, 9637–9644. [CrossRef]

153. Schmidt, C.; Jones, M.J.; Ulrich, J. The Influence of Additives and Impurities on Crystallization. In *Crystallization*; John Wiley & Sons, Ltd.: Hoboken, NJ, USA, 2013; pp. 105–127.

154. Liu, D.; Liu, Y.; Dai, F.; Zhao, J.; Yang, K.; Liu, C. Size- and morphology-controllable synthesis of MIL-96 (Al) by hydrolysis and coordination modulation of dual aluminium source and ligand systems. *Dalton Trans.* **2015**, *44*, 16421–16429. [CrossRef]

155. Liu, D.; Yan, L.; Li, L.; Gu, X.; Dai, P.; Yang, L.; Liu, Y.; Liu, C.; Zhao, G.; Zhao, X. Impact of moderative ligand hydrolysis on morphology evolution and the morphology-dependent breathing effect performance of MIL-53(Al). *CrystEngComm* **2018**, *20*, 2102–2111. [CrossRef]

156. Zelinskii, G.E.; Belov, A.S.; Lebed, E.G.; Vologzhanina, A.V.; Novikov, V.V.; Voloshin, Y.Z. Synthesis, structure and reactivity of iron(II) clathrochelates with terminal formyl (acetal) groups. *Inorg. Chim. Acta* **2016**, *440*, 154–164. [CrossRef]

157. Koradia, V.; de Diego, H.L.; Elema, M.R.; Rantanen, J. Integrated Approach to Study the Dehydration Kinetics of Nitrofurantoin Monohydrate. *J. Pharm. Sci.* **2010**, *99*, 3966–3976. [CrossRef]

158. Thakuria, R.; Eddleston, M.D.; Chow, E.H.H.; Lloyd, G.O.; Aldous, B.J.; Krzyzaniak, J.F.; Bond, A.D.; Jones, W. Use of in Situ Atomic Force Microscopy to Follow Phase Changes at Crystal Surfaces in Real Time. *Angew. Chem. Int. Ed.* **2013**, *52*, 10541–10544. [CrossRef]

159. Park, Y.; Boerrigter, S.X.M.; Yeon, J.; Lee, S.H.; Kang, S.K.; Lee, E.H. New Metastable Packing Polymorph of Donepezil Grown on Stable Polymorph Substrates. *Cryst. Growth Des.* **2016**, *16*, 2552–2560. [CrossRef]

160. Bond, A.D.; Boese, R.; Desiraju, G.R. On the Polymorphism of Aspirin: Crystalline Aspirin as Intergrowths of Two "Polymorphic" Domains. *Angew. Chem. Int. Ed.* **2007**, *46*, 618–622. [CrossRef]

161. Mishra, M.K.; Desiraju, G.R.; Ramamurty, U.; Bond, A.D. Studying Microstructure in Molecular Crystals with Nanoindentation: Intergrowth Polymorphism in Felodipine. *Angew. Chem. Int. Ed.* **2014**, *53*, 13102–13105. [CrossRef]

162. Croker, D.M.; Davey, R.J.; Rasmuson, Å.C.; Seaton, C.C. Solution mediated phase transformations between co-crystals. *CrystEngComm* **2013**, *15*, 2044–2047. [CrossRef]

163. Bhandary, S.; Chopra, D. Silicone Oil Induced Spontaneous Single-Crystal-to-Single-Crystal Phase Transitions in Ethynyl Substituted ortho- and meta-Fluorinated Benzamides. *Cryst. Growth Des.* **2017**, *17*, 4533–4540. [CrossRef]

164. Ravi, A.; Sureshan, K.M. Tunable Mechanical Response from a Crystal Undergoing Topochemical Dimerization: Instant Explosion at a Faster Rate and Chemical Storage of a Harvestable Explosion at a Slower Rate. *Angew. Chem. Int. Ed.* **2018**, *57*, 9362–9366. [CrossRef]

165. Lee, A.Y.; Ulman, A.; Myerson, A.S. Crystallization of Amino Acids on Self-Assembled Monolayers of Rigid Thiols on Gold. *Langmuir* **2002**, *18*, 5886–5898. [CrossRef]

166. Cox, J.R.; Dabros, M.; Shaffer, J.A.; Thalladi, V.R. Selective Crystal Growth of the Anhydrous and Monohydrate Forms of Theophylline on Self-Assembled Monolayers. *Angew. Chem. Int. Ed.* **2007**, *46*, 1988–1991. [CrossRef]

167. Urbelis, J.H.; Swift, J.A. Phase-Selective Crystallization of Perylene on Monolayer Templates. *Cryst. Growth Des.* **2014**, *14*, 5244–5251. [CrossRef]

168. Zhang, J.; Liu, A.; Han, Y.; Ren, Y.; Gong, J.; Li, W.; Wang, J. Effects of Self-Assembled Monolayers on Selective Crystallization of Tolbutamide. *Cryst. Growth Des.* **2011**, *11*, 5498–5506. [CrossRef]

169. Solomos, M.A.; Capacci-Daniel, C.; Rubinson, J.F.; Swift, J.A. Polymorph Selection via Sublimation onto Siloxane Templates. *Cryst. Growth Des.* **2018**, *18*, 6965–6972. [CrossRef]

170. Serrano, D.R.; Mugheirbi, N.A.; O'Connell, P.; Leddy, N.; Healy, A.M.; Tajber, L. Impact of Substrate Properties on the Formation of Spherulitic Films: A Case Study of Salbutamol Sulfate. *Cryst. Growth Des.* **2016**, *16*, 3853–3858. [CrossRef]

171. Patel, M.A.; Nguyen, B.; Chadwick, K. Predicting the Nucleation Induction Time Based on Preferred Intermolecular Interactions. *Cryst. Growth Des.* **2017**, *17*, 4613–4621. [CrossRef]

172. Telford, R.; Seaton, C.C.; Clout, A.; Buanz, A.; Gaisford, S.; Williams, G.R.; Prior, T.J.; Okoye, C.H.; Munshi, T.; Scowen, I.J. Stabilisation of metastable polymorphs: The case of paracetamol form III. *Chem. Commun.* **2016**, *52*, 12028–12031. [CrossRef]

173. Zhang, K.; Xu, S.; Liu, S.; Tang, W.; Fu, X.; Gong, J. Novel Strategy to Control Polymorph Nucleation of Gamma Pyrazinamide by Preferred Intermolecular Interactions during Heterogeneous Nucleation. *Cryst. Growth Des.* **2018**, *18*, 4874–4879. [CrossRef]

174. Rao, K.P.; Higuchi, M.; Sumida, K.; Furukawa, S.; Duan, J.; Kitagawa, S. Design of Superhydrophobic Porous Coordination Polymers through the Introduction of External Surface Corrugation by the Use of an Aromatic Hydrocarbon Building Unit. *Angew. Chem. Int. Ed.* **2014**, *53*, 8225–8230. [CrossRef]

175. Zhang, J.; Mitchell, L.A.; Parrish, D.A.; Shreeve, J.M. Enforced Layer-by-Layer Stacking of Energetic Salts towards High-Performance Insensitive Energetic Materials. *J. Am. Chem. Soc.* **2015**, *137*, 10532–10535. [CrossRef]

176. Wang, K.; Mishra, M.K.; Sun, C.C. Exceptionally Elastic Single-Component Pharmaceutical Crystals. *Chem. Mater.* **2019**, *31*, 1794–1799. [CrossRef]

177. Devarapalli, R.; Kadambi, S.B.; Chen, C.-T.; Krishna, G.R.; Kammari, B.R.; Buehler, M.J.; Ramamurty, U.; Reddy, C.M. Remarkably Distinct Mechanical Flexibility in Three Structurally Similar Semiconducting Organic Crystals Studied by Nanoindentation and Molecular Dynamics. *Chem. Mater.* **2019**, *31*, 1391–1402. [CrossRef]

178. SeethaLekshmi, S.; Kiran, M.S.R.N.; Ramamurty, U.; Varughese, S. Molecular Basis for the Mechanical Response of Sulfa Drug Crystals. *Chem.—Eur. J.* **2019**, *25*, 526–537. [CrossRef] [PubMed]

179. Krishna, G.R.; Shi, L.; Bag, P.P.; Sun, C.C.; Reddy, C.M. Correlation Among Crystal Structure, Mechanical Behavior, and Tabletability in the Co-Crystals of Vanillin Isomers. *Cryst. Growth Des.* **2015**, *15*, 1827–1832. [CrossRef]

180. Zolotarev, P.N.; Moret, M.; Rizzato, S.; Proserpio, D.M. Searching New Crystalline Substrates for OMBE: Topological and Energetic Aspects of Cleavable Organic Crystals. *Cryst. Growth Des.* **2016**, *16*, 1572–1582. [CrossRef]

181. Reddy, C.M.; Gundakaram, R.C.; Basavoju, S.; Kirchner, M.T.; Padmanabhan, K.A.; Desiraju, G.R. Structural basis for bending of organic crystals. *Chem. Commun.* **2005**, 3945–3947. [CrossRef] [PubMed]

182. Chattoraj, S.; Shi, L.; Sun, C.C. Understanding the relationship between crystal structure, plasticity and compaction behaviour of theophylline, methyl gallate, and their 1:1 co-crystal. *CrystEngComm* **2010**, *12*, 2466–2472. [CrossRef]

183. Veits, G.K.; Carter, K.K.; Cox, S.J.; McNeil, A.J. Developing a Gel-Based Sensor Using Crystal Morphology Prediction. *J. Am. Chem. Soc.* **2016**, *138*, 12228–12233. [CrossRef] [PubMed]

184. Bučar, D.-K.; Lancaster, R.W.; Bernstein, J. Disappearing Polymorphs Revisited. *Angew. Chem. Int. Ed.* **2015**, *54*, 6972–6993. [CrossRef] [PubMed]

185. Galek, P.T.A.; Allen, F.H.; Fábián, L.; Feeder, N. Knowledge-based H-bond prediction to aid experimental polymorph screening. *CrystEngComm* **2009**, *11*, 2634–2639. [CrossRef]

186. Galek, P.T.A.; Fábián, L.; Motherwell, W.D.S.; Allen, F.H.; Feeder, N. Knowledge-based model of hydrogen-bonding propensity in organic crystals. *Acta Cryst. Sect. B* **2007**, *63*, 768–782. [CrossRef]

187. Galek, P.T.A.; Chisholm, J.A.; Pidcock, E.; Wood, P.A. Hydrogen-bond coordination in organic crystal structures: Statistics, predictions and applications. *Acta Cryst. Sect. B* **2014**, *70*, 91–105. [CrossRef]

188. Galek, P.T.A.; Pidcock, E.; Wood, P.A.; Bruno, I.J.; Groom, C.R. One in half a million: A solid form informatics study of a pharmaceutical crystal structure. *CrystEngComm* **2012**, *14*, 2391–2403. [CrossRef]

189. Abramov, Y.A. Current Computational Approaches to Support Pharmaceutical Solid Form Selection. *Org. Process Res. Dev.* **2013**, *17*, 472–485. [CrossRef]

190. Feeder, N.; Pidcock, E.; Reilly, A.M.; Sadiq, G.; Doherty, C.L.; Back, K.R.; Meenan, P.; Docherty, R. The integration of solid-form informatics into solid-form selection. *J. Pharm. Pharm.* **2015**, *67*, 857–868. [CrossRef] [PubMed]

191. Allen, F.H.; Wood, P.A.; Galek, P.T.A. The versatile role of the ethynyl group in crystal packing: An interaction propensity study. *Acta Cryst. Sect. B* **2013**, *69*, 281–287. [CrossRef] [PubMed]

192. Allen, F.H.; Wood, P.A.; Galek, P.T.A. Role of chloroform and dichloromethane solvent molecules in crystal packing: An interaction propensity study. *Acta Cryst. Sect. B* **2013**, *69*, 379–388. [CrossRef] [PubMed]

193. Srirambhatla, V.K.; Kraft, A.; Watt, S.; Powell, A.V. Crystal Design Approaches for the Synthesis of Paracetamol Co-Crystals. *Cryst. Growth Des.* **2012**, *12*, 4870–4879. [CrossRef]

194. Vologzhanina, A.V.; Sokolov, A.V.; Purygin, P.P.; Zolotarev, P.N.; Blatov, V.A. Knowledge-Based Approaches to H-Bonding Patterns in Heterocycle-1-Carbohydrazoneamides. *Cryst. Growth Des.* **2016**, *16*, 6354–6362. [CrossRef]

195. Hulme, A.T.; Johnston, A.; Florence, A.J.; Fernandes, P.; Shankland, K.; Bedford, C.T.; Welch, G.W.A.; Sadiq, G.; Haynes, D.A.; Motherwell, W.D.S.; et al. Search for a Predicted Hydrogen Bonding Motif—A Multidisciplinary Investigation into the Polymorphism of 3-Azabicyclo[3.3.1]nonane-2,4-dione. *J. Am. Chem. Soc.* **2007**, *129*, 3649–3657. [CrossRef] [PubMed]

196. Nauha, E.; Bernstein, J. "Predicting" Polymorphs of Pharmaceuticals Using Hydrogen Bond Propensities: Probenecid and Its Two Single-Crystal-to-Single-Crystal Phase Transitions. *J. Pharm. Sci.* **2015**, *104*, 2056–2061. [CrossRef]

197. Nauha, E.; Bernstein, J. "Predicting" Crystal Forms of Pharmaceuticals Using Hydrogen Bond Propensities: Two Test Cases. *Cryst. Growth Des.* **2014**, *14*, 4364–4370. [CrossRef]

198. Jones, W.; Motherwell, W.D.S.; Trask, A.V. Pharmaceutical Cocrystals: An Emerging Approach to Physical Property Enhancement. *MRS Bull.* **2006**, *31*, 875–879. [CrossRef]

199. Sathisaran, I.; Dalvi, S.V. Engineering Cocrystals of Poorly Water-Soluble Drugs to Enhance Dissolution in Aqueous Medium. *Pharmaceutics* **2018**, *10*, 108. [CrossRef] [PubMed]

200. Schultheiss, N.; Newman, A. Pharmaceutical Cocrystals and Their Physicochemical Properties. *Cryst. Growth Des.* **2009**, *9*, 2950–2967. [CrossRef] [PubMed]

201. Karki, S.; Friščić, T.; Fábián, L.; Laity, P.R.; Day, G.M.; Jones, W. Improving Mechanical Properties of Crystalline Solids by Cocrystal Formation: New Compressible Forms of Paracetamol. *Adv. Mater.* **2009**, *21*, 3905–3909. [CrossRef]

202. Saha, S.; Desiraju, G.R. Acid···Amide Supramolecular Synthon in Cocrystals: From Spectroscopic Detection to Property Engineering. *J. Am. Chem. Soc.* **2018**, *140*, 6361–6373. [CrossRef] [PubMed]

203. Zhang, J.; Shreeve, J.M. Time for pairing: Cocrystals as advanced energetic materials. *CrystEngComm* **2016**, *18*, 6124–6133. [CrossRef]

204. Bolton, O.; Simke, L.R.; Pagoria, P.F.; Matzger, A.J. High Power Explosive with Good Sensitivity: A 2:1 Cocrystal of CL-20: HMX. *Cryst. Growth Des.* **2012**, *12*, 4311–4314. [CrossRef]

205. Gryl, M.; Seidler, T.; Stadnicka, K.; Matulková, I.; Němec, I.; Tesařová, N.; Němec, P. The crystal structure and optical properties of a pharmaceutical co-crystal—The case of the melamine–barbital addition compound. *CrystEngComm* **2014**, *16*, 5765–5768. [CrossRef]

206. Christopherson, J.-C.; Topić, F.; Barrett, C.J.; Friščić, T. Halogen-Bonded Cocrystals as Optical Materials: Next-Generation Control over Light–Matter Interactions. *Cryst. Growth Des.* **2018**, *18*, 1245–1259. [CrossRef]

207. Noa, F.M.A.; Mehlana, G. Co-crystals and salts of vanillic acid and vanillin with amines. *CrystEngComm* **2018**, *20*, 896–905.

208. Walsh, R.D.B.; Bradner, M.W.; Fleischman, S.; Morales, L.A.; Moulton, B.; Rodríguez-Hornedo, N.; Zaworotko, M.J. Crystal engineering of the composition of pharmaceutical phases. *Chem. Commun.* **2003**, 186–187. [CrossRef] [PubMed]

209. Price(Sally), S.L. Computed Crystal Energy Landscapes for Understanding and Predicting Organic Crystal Structures and Polymorphism. *Acc. Chem. Res.* **2009**, *42*, 117–126.

210. Etter, M.C. Encoding and decoding hydrogen-bond patterns of organic compounds. *Acc. Chem. Res.* **1990**, *23*, 120–126. [CrossRef]

211. Delori, A.; Galek, P.T.A.; Pidcock, E.; Jones, W. Quantifying Homo- and Heteromolecular Hydrogen Bonds as a Guide for Adduct Formation. *Chem. Eur. J.* **2012**, *18*, 6835–6846. [CrossRef] [PubMed]

212. Delori, A.; Galek, P.T.A.; Pidcock, E.; Patni, M.; Jones, W. Knowledge-based hydrogen bond prediction and the synthesis of salts and cocrystals of the anti-malarial drug pyrimethamine with various drug and GRAS molecules. *CrystEngComm* **2013**, *15*, 2916–2928. [CrossRef]

213. Bhogala, B.R.; Basavoju, S.; Nangia, A. Tape and layer structures in cocrystals of some di- and tricarboxylic acids with 4,4'-bipyridines and isonicotinamide. From binary to ternary cocrystals. *CrystEngComm* **2005**, *7*, 551–562. [CrossRef]

214. Cruz-Cabeza, A.J. Acid–base crystalline complexes and the pKa rule. *CrystEngComm* **2012**, *14*, 6362–6365. [CrossRef]

215. Chamorro Orué, A.I.; Boese, R.; Schauerte, C.; Merz, K. An Experimental and Theoretical Approach to Control Salt vs Cocrystal vs Hybrid Formation—Crystal Engineering of an E/Z-Butenedioic Acid/Phthalazine System. *Cryst. Growth Des.* **2019**, *19*, 1616–1620. [CrossRef]

216. Sandhu, B.; McLean, A.; Sinha, A.S.; Desper, J.; Sarjeant, A.A.; Vyas, S.; Reutzel-Edens, S.M.; Aakeröy, C.B. Evaluating Competing Intermolecular Interactions through Molecular Electrostatic Potentials and Hydrogen-Bond Propensities. *Cryst. Growth Des.* **2018**, *18*, 466–478. [CrossRef]

217. Wang, J.-R.; Ye, C.; Mei, X. Structural and physicochemical aspects of hydrochlorothiazide co-crystals. *CrystEngComm* **2014**, *16*, 6996–7003. [CrossRef]

218. Eddleston, M.D.; Arhangelskis, M.; Fábián, L.; Tizzard, G.J.; Coles, S.J.; Jones, W. Investigation of an Amide-Pseudo Amide Hydrogen Bonding Motif within a Series of Theophylline: Amide Cocrystals. *Cryst. Growth Des.* **2016**, *16*, 51–58. [CrossRef]

219. Corpinot, M.K.; Stratford, S.A.; Arhangelskis, M.; Anka-Lufford, J.; Halasz, I.; Judaš, N.; Jones, W.; Bučar, D.-K. On the predictability of supramolecular interactions in molecular cocrystals—the view from the bench. *CrystEngComm* **2016**, *18*, 5434–5439. [CrossRef]

220. Wang, T.; Stevens, J.S.; Vetter, T.; Whitehead, G.F.S.; Vitorica-Yrezabal, I.J.; Hao, H.; Cruz-Cabeza, A.J. Salts, Cocrystals, and Ionic Cocrystals of a "Simple" Tautomeric Compound. *Cryst. Growth Des.* **2018**, *18*, 6973–6983. [CrossRef]

221. Mapp, L.K.; Coles, S.J.; Aitipamula, S. Design of Cocrystals for Molecules with Limited Hydrogen Bonding Functionalities: Propyphenazone as a Model System. *Cryst. Growth Des.* **2017**, *17*, 163–174. [CrossRef]

222. Liu, F.; Song, Y.; Liu, Y.-N.; Li, Y.-T.; Wu, Z.-Y.; Yan, C.-W. Drug-Bridge-Drug Ternary Cocrystallization Strategy for Antituberculosis Drugs Combination. *Cryst. Growth Des.* **2018**, *18*, 1283–1286. [CrossRef]

223. Almansa, C.; Mercè, R.; Tesson, N.; Farran, J.; Tomàs, J.; Plata-Salamán, C.R. Co-crystal of Tramadol Hydrochloride–Celecoxib (ctc): A Novel API–API Co-crystal for the Treatment of Pain. *Cryst. Growth Des.* **2017**, *17*, 1884–1892. [CrossRef]

224. Skořepová, E.; Hušák, M.; Čejka, J.; Zámostný, P.; Kratochvíl, B. Increasing dissolution of trospium chloride by co-crystallization with urea. *J. Cryst. Growth* **2014**, *399*, 19–26. [CrossRef]

225. Chandra, A.; Ghate, M.V.; Aithal, K.S.; Lewis, S.A. In silico prediction coupled with in vitro experiments and absorption modeling to study the inclusion complex of telmisartan with modified beta-cyclodextrin. *J. Incl. Phenom. Macrocycl. Chem.* **2018**, *91*, 47–60. [CrossRef]

226. Tothadi, S.; Desiraju, G.R. Designing ternary cocrystals with hydrogen bonds and halogen bonds. *Chem. Commun.* **2013**, *49*, 7791–7793. [CrossRef]

227. Arman, H.D.; Gieseking, R.L.; Hanks, T.W.; Pennington, W.T. Complementary halogen and hydrogen bonding: Sulfur···iodine interactions and thioamide ribbons. *Chem. Commun.* **2010**, *46*, 1854–1856. [CrossRef]

228. Topić, F.; Rissanen, K. Systematic Construction of Ternary Cocrystals by Orthogonal and Robust Hydrogen and Halogen Bonds. *J. Am. Chem. Soc.* **2016**, *138*, 6610–6616. [CrossRef] [PubMed]

229. Seaton, C.C.; Blagden, N.; Munshi, T.; Scowen, I.J. Creation of Ternary Multicomponent Crystals by Exploitation of Charge-Transfer Interactions. *Chem.—Eur. J.* **2013**, *19*, 10663–10671. [CrossRef] [PubMed]

230. Thomas, L.H.; Blagden, N.; Gutmann, M.J.; Kallay, A.A.; Parkin, A.; Seaton, C.C.; Wilson, C.C. Tuning Proton Behavior in a Ternary Molecular Complex. *Cryst. Growth Des.* **2010**, *10*, 2770–2774. [CrossRef]

231. Mishra, M.K.; Mukherjee, A.; Ramamurty, U.; Desiraju, G.R. Crystal chemistry and photomechanical behavior of 3,4-dimethoxycinnamic acid: Correlation between maximum yield in the solid-state topochemical reaction and cooperative molecular motion. *IUCrJ* **2015**, *2*, 653–660. [CrossRef] [PubMed]

232. Mir, N.A.; Dubey, R.; Desiraju, G.R. Four- and five-component molecular solids: Crystal engineering strategies based on structural inequivalence. *IUCrJ* **2016**, *3*, 96–101. [CrossRef] [PubMed]

233. Paul, M.; Chakraborty, S.; Desiraju, G.R. Six-Component Molecular Solids: ABC[D$_{1-(x+y)}$E$_x$F$_y$]$_2$. *J. Am. Chem. Soc.* **2018**, *140*, 2309–2315. [CrossRef]

234. Sander, J.R.G.; Bučar, D.-K.; Henry, R.F.; Giangiorgi, B.N.; Zhang, G.G.Z.; MacGillivray, L.R. 'Masked synthons' in crystal engineering: Insulated components in acetaminophen cocrystal hydrates. *CrystEngComm* **2013**, *15*, 4816–4822. [CrossRef]

235. Oswald, I.D.H.; Allan, D.R.; McGregor, P.A.; Motherwell, W.D.S.; Parsons, S.; Pulham, C.R. The formation of paracetamol (acetaminophen) adducts with hydrogen-bond acceptors. *Acta Cryst. Sect. B* **2002**, *58*, 1057–1066. [CrossRef]

236. Oswald, I.D.H.; Motherwell, W.D.S.; Parsons, S.; Pidcock, E.; Pulham, C.R. Rationalisation of Co-Crystal Formation Through Knowledge-Mining. *Crystallogr. Rev.* **2004**, *10*, 57–66. [CrossRef]

237. Wicker, J.G.P.; Crowley, L.M.; Robshaw, O.; Little, E.J.; Stokes, S.P.; Cooper, R.I.; Lawrence, S.E. Will they co-crystallize? *CrystEngComm* **2017**, *19*, 5336–5340. [CrossRef]

238. Fábián, L. Cambridge Structural Database Analysis of Molecular Complementarity in Cocrystals. *Cryst. Growth Des.* **2009**, *9*, 1436–1443. [CrossRef]

239. Surov, A.O.; Churakov, A.V.; Proshin, A.N.; Dai, X.-L.; Lu, T.; Perlovich, G.L. Cocrystals of a 1,2,4-thiadiazole-based potent neuroprotector with gallic acid: Solubility, thermodynamic stability relationships and formation pathways. *Phys. Chem. Chem. Phys.* **2018**, *20*, 14469–14481. [CrossRef] [PubMed]

240. Karki, S.; Friščić, T.; Fábián, L.; Jones, W. New solid forms of artemisinin obtained through cocrystallisation. *CrystEngComm* **2010**, *12*, 4038–4041. [CrossRef]

241. Alsubaie, M.; Aljohani, M.; Erxleben, A.; McArdle, P. Cocrystal Forms of the BCS Class IV Drug Sulfamethoxazole. *Cryst. Growth Des.* **2018**, *18*, 3902–3912. [CrossRef]

242. Cadden, J.; Klooster, W.T.; Coles, S.J.; Aitipamula, S. Cocrystals of Leflunomide: Design, Structural, and Physicochemical Evaluation. *Cryst. Growth Des.* **2019**, *19*, 3923–3933. [CrossRef]

243. Gryl, M.; Rydz, A.; Wojnarska, J.; Krawczuk, A.; Kozieł, M.; Seidler, T.; Ostrowska, K.; Marzec, M.; Stadnicka, K.M. Origin of chromic effects and crystal-to-crystal phase transition in the polymorphs of tyraminium violurate. *IUCrJ* **2019**, *6*, 226–237. [CrossRef] [PubMed]

244. Robertson, C.C.; Wright, J.S.; Carrington, E.J.; Perutz, R.N.; Hunter, C.A.; Brammer, L. Hydrogen bonding vs. halogen bonding: The solvent decides. *Chem. Sci.* **2017**, *8*, 5392–5398. [CrossRef]

245. Wood, P.A.; Olsson, T.S.G.; Cole, J.C.; Cottrell, S.J.; Feeder, N.; Galek, P.T.A.; Groom, C.R.; Pidcock, E. Evaluation of molecular crystal structures using Full Interaction Maps. *CrystEngComm* **2013**, *15*, 65–72. [CrossRef]

246. Golovanov, A.A.; Latypova, D.R.; Bekin, V.V.; Pisareva, V.S.; Vologzhanina, A.V.; Dokichev, V.A. Synthesis of 1,5-disubstituted (E)-pent-2-en-4-yn-1-ones. *Russ. J. Org. Chem.* **2013**, *49*, 1264–1269. [CrossRef]

247. Vologzhanina, A.V.; Golovanov, A.A.; Gusev, D.M.; Odin, I.S.; Apreyan, R.A.; Suponitsky, K.Yu. Intermolecular Interactions and Second-Harmonic Generation Properties of (E)-1,5-Diarylpentenyn-1-ones. *Cryst. Growth Des.* **2014**, *14*, 4402–4410. [CrossRef]

248. Mugheirbi, N.A.; Tajber, L. Crystal Habits of Itraconazole Microcrystals: Unusual Isomorphic Intergrowths Induced via Tuning Recrystallization Conditions. *Mol. Pharm.* **2015**, *12*, 3468–3478. [CrossRef] [PubMed]

249. Sandhu, B.; Sinha, A.S.; Desper, J.; Aakeröy, C.B. Modulating the physical properties of solid forms of urea using co-crystallization technology. *Chem. Commun.* **2018**, *54*, 4657–4660. [CrossRef] [PubMed]

250. Wojnarska, J.; Gryl, M.; Seidler, T.; Stadnicka, K.M. Crystal engineering, optical properties and electron density distribution of polar multicomponent materials containing sulfanilamide. *CrystEngComm* **2018**, *20*, 3638–3646. [CrossRef]

251. Xing, G.; Bassanetti, I.; Ben, T.; Bracco, S.; Sozzani, P.; Marchiò, L.; Comotti, A. Multifunctional Organosulfonate Anions Self-Assembled with Organic Cations by Charge-Assisted Hydrogen Bonds and the Cooperation of Water. *Cryst. Growth Des.* **2018**, *18*, 2082–2092. [CrossRef]

252. Honorato, J.; Colina-Vegas, L.; Correa, R.S.; Guedes, A.P.M.; Miyata, M.; Pavan, F.R.; Ellena, J.; Batista, A.A. Esterification of the free carboxylic group from the lutidinic acid ligand as a tool to improve the cytotoxicity of Ru(II) complexes. *Inorg. Chem. Front.* **2019**, *6*, 376–390. [CrossRef]

253. Kodrin, I.; Soldin, Ž.; Aakeröy, C.B.; Đaković, M. Role of the "Weakest Link" in a Pressure-Driven Phase Transition of Two Polytypic Polymorphs. *Cryst. Growth Des.* **2016**, *16*, 2040–2051. [CrossRef]

254. Moggach, S.A.; Marshall, W.G.; Rogers, D.M.; Parsons, S. How focussing on hydrogen bonding interactions in amino acids can miss the bigger picture: A high-pressure neutron powder diffraction study of ε-glycine. *CrystEngComm* **2015**, *17*, 5315–5328. [CrossRef]

255. Voronova, E.D.; Golovanov, A.A.; Suponitsky, K.Yu.; Fedyanin, I.V.; Vologzhanina, A.V. Theoretical Charge Density Analysis and Nonlinear Optical Properties of Quasi-Planar 1-Aryl(hetaryl)-5-phenylpent-1-en-4-yn-3-ones. *Cryst. Growth Des.* **2016**, *16*, 3859–3868. [CrossRef]

256. Vráblová, A.; Černák, J.; Falvello, L.R.; Tomás, M. Polymorphism of the dinuclear Co^{III}–Schiff base complex [Co^2(o-van-en)^3]·4CH^{3}CN (o-van-en is a salen-type ligand). *Acta Cryst. Sect. C* **2019**, *75*, 433–442. [CrossRef]

257. Voronova, E.D.; Golovanov, A.A.; Odin, I.S.; Anisimov, M.A.; Dorovatovskii, P.V.; Zubavichus, Y.V.; Vologzhanina, A.V. Peculiarities of supramolecular organization of cyclic ketones with vinylacetylene fragments. *Acta Cryst. Sect. C* **2018**, *74*, 1674–1683. [CrossRef]

258. Dotsenko, V.V.; Frolov, K.A.; Krivokolysko, S.G.; Chigorina, E.A.; Pekhtereva, T.M.; Suykov, S.Yu.; Papayanina, E.S.; Dmitrienko, A.O.; Bushmarinov, I.S. Aminomethylation of morpholinium and N-methylmorpholinium 3,5-dicyano-4,4-dimethyl-6-oxo-1,4,5,6-tetrahydropyridine-2-thiolates. *Chem. Heterocycl. Comp.* **2016**, *52*, 116–127. [CrossRef]

259. Carletta, A.; Zbačnik, M.; Van Gysel, M.; Vitković, M.; Tumanov, N.; Stilinović, V.; Wouters, J.; Cinčić, D. Playing with Isomerism: Cocrystallization of Isomeric N-Salicylideneaminopyridines with Perfluorinated Compounds as Halogen Bond Donors and Its Impact on Photochromism. *Cryst. Growth Des.* **2018**, *18*, 6833–6842. [CrossRef]

260. Mugheirbi, N.A.; Tajber, L. Mesophase and size manipulation of itraconazole liquid crystalline nanoparticles produced via quasi nanoemulsion precipitation. *Eur. J. Pharm. Biopharm.* **2015**, *96*, 226–236. [CrossRef] [PubMed]

261. Cole, J.C.; Kabova, E.A.; Shankland, K. Utilizing organic and organometallic structural data in powder diffraction. *Powder Diffr.* **2014**, *29*, S19–S30. [CrossRef]

262. Anisimov, A.A.; Zhemchugov, P.V.; Milenin, S.A.; Goloveshkin, A.S.; Tsareva, U.S.; Bushmarinov, I.S.; Korlyukov, A.A.; Takazova, R.U.; Molodtsova, Y.A.; Muzafarov, A.M.; et al. Sodium cis-tetratolylcyclotetrasiloxanolate and cis-tritolylcyclotrisiloxanolate: Synthesis, structure and their mutual transformations. *J. Organomet. Chem.* **2016**, *823*, 103–111. [CrossRef]

263. Taylor, R.; Cole, J.; Korb, O.; McCabe, P. Knowledge-Based Libraries for Predicting the Geometric Preferences of Druglike Molecules. *J. Chem. Inf. Model.* **2014**, *54*, 2500–2514. [CrossRef] [PubMed]

264. Cole, J.C.; Korb, O.; McCabe, P.; Read, M.G.; Taylor, R. Knowledge-Based Conformer Generation Using the Cambridge Structural Database. *J. Chem. Inf. Model.* **2018**, *58*, 615–629. [CrossRef] [PubMed]

265. Giangreco, I.; Olsson, T.S.G.; Cole, J.C.; Packer, M.J. Assessment of a Cambridge Structural Database-Driven Overlay Program. *J. Chem. Inf. Model.* **2014**, *54*, 3091–3098. [CrossRef] [PubMed]

266. Capucci, D.; Balestri, D.; Mazzeo, P.P.; Pelagatti, P.; Rubini, K.; Bacchi, A. Liquid Nicotine Tamed in Solid Forms by Cocrystallization. *Cryst. Growth Des.* **2017**, *17*, 4958–4964. [CrossRef]

267. Habgood, M. Bioactive focus in conformational ensembles: A pluralistic approach. *J. Comput. Aided Mol. Des.* **2017**, *31*, 1073–1083. [CrossRef] [PubMed]

268. Iuzzolino, L.; Reilly, A.M.; McCabe, P.; Price, S.L. Use of Crystal Structure Informatics for Defining the Conformational Space Needed for Predicting Crystal Structures of Pharmaceutical Molecules. *J. Chem. Theory Comput.* **2017**, *13*, 5163–5171. [CrossRef]

269. Iuzzolino, L.; McCabe, P.; Price, S.L.; Brandenburg, J.G. Crystal structure prediction of flexible pharmaceutical-like molecules: Density functional tight-binding as an intermediate optimisation method and for free energy estimation. *Faraday Discuss.* **2018**, *211*, 275–296. [CrossRef] [PubMed]

270. Pal, S.K.; Zewail, A.H. Dynamics of Water in Biological Recognition. *Chem. Rev.* **2004**, *104*, 2099–2124. [CrossRef]

271. Terao, H.; Sugawara, T.; Kita, Y.; Sato, N.; Kaho, E.; Takeda, S. Proton Relay in a One-Dimensional Hydrogen-Bonded Chain Composed of Water Molecules and a Squaric Acid Derivative. *J. Am. Chem. Soc.* **2001**, *123*, 10468–10474. [CrossRef] [PubMed]

272. Ananyev, I.V.; Bushmarinov, I.S.; Ushakov, I.E.; Aitkulova, A.I.; Lyssenko, K.A. Tuning of the double-well potential of short strong hydrogen bonds by ionic interactions in alkali metal hydrodicarboxylates. *RSC Adv.* **2015**, *5*, 97495–97502. [CrossRef]

273. Hickey, M.B.; Peterson, M.L.; Manas, E.S.; Alvarez, J.; Haeffner, F.; Almarsson, Ö. Hydrates and Solid-State Reactivity: A Survey of β-Lactam Antibiotics. *J. Pharm. Sci.* **2007**, *96*, 1090–1099. [CrossRef]

274. Xu, H.-R.; Zhang, Q.-C.; Ren, Y.-P.; Zhao, H.-X.; Long, L.-S.; Huang, R.-B.; Zheng, L.-S. The influence of water on dielectric property in cocrystal compound of [orotic acid] [melamine]·H₂O. *CrystEngComm* **2011**, *13*, 6361–6364. [CrossRef]

275. Ananyev, I.V.; Barzilovich, P.Yu.; Lyssenko, K.A. Evidence for the Zundel-like Character of Oxoethylidenediphosphonic Acid Hydrate. *Mendeleev Commun.* **2012**, *22*, 242–244. [CrossRef]

276. Liu, F.; Hooks, D.E.; Li, N.; Mara, N.A.; Swift, J.A. Mechanical Properties of Anhydrous and Hydrated Uric Acid Crystals. *Chem. Mater.* **2018**, *30*, 3798–3805. [CrossRef]

277. Silverstein, K.A.T.; Haymet, A.D.J.; Dill, K.A. A Simple Model of Water and the Hydrophobic Effect. *J. Am. Chem. Soc.* **1998**, *120*, 3166–3175. [CrossRef]

278. Ludwig, R. Water: From Clusters to the Bulk. *Angew. Chem. Int. Ed.* **2001**, *40*, 1808–1827. [CrossRef]

279. Takeuchi, F.; Hiratsuka, M.; Ohmura, R.; Alavi, S.; Sum, A.K.; Yasuoka, K. Water proton configurations in structures I, II, and H clathrate hydrate unit cells. *J. Chem. Phys.* **2013**, *138*, 124504. [CrossRef] [PubMed]

280. Joseph, A.; Alves, J.S.R.; Bernardes, C.E.S.; Piedade, M.F.M.; Piedade, M.E.M. da Tautomer selection through solvate formation: The case of 5-hydroxynicotinic acid. *CrystEngComm* **2019**, *21*, 2220–2233. [CrossRef]

281. Medvedev, A.G.; Mikhailov, A.A.; Prikhodchenko, P.V.; Tripol'skaya, T.A.; Lev, O.; Churakov, A.V. Crystal structures of pyridinemonocarboxylic acid peroxosolvates. *Russ. Chem. Bull.* **2013**, *62*, 1871–1876. [CrossRef]

282. Gillon, A.L.; Feeder, N.; Davey, R.J.; Storey, R. Hydration in Molecular Crystals. A Cambridge Structural Database Analysis. *Cryst. Growth Des.* **2003**, *3*, 663–673. [CrossRef]

283. Puntus, L.N.; Lyssenko, K.A.; Pekareva, I.S.; Bünzli, J.-C.G. Intermolecular Interactions as Actors in Energy-Transfer Processes in Lanthanide Complexes with 2,2'-Bipyridine. *J. Phys. Chem. B* **2009**, *113*, 9265–9277. [CrossRef]

284. Nelyubina, Y.V.; Puntus, L.N.; Lyssenko, K.A. The Dark Side of Hydrogen Bonds in the Design of Optical Materials: A Charge-Density Perspective. *Chem. Eur. J.* **2014**, *20*, 2860–2865. [CrossRef]

285. Banaru, A.M.; Slovokhotov, Y.L. Crystal hydrates of organic compounds. *J. Struct. Chem.* **2015**, *56*, 967–982. [CrossRef]

286. Dobrzycki, Ł.; Socha, P.; Ciesielski, A.; Boese, R.; Cyrański, M.K. Formation of Crystalline Hydrates by Nonionic Chaotropes and Kosmotropes: Case of Piperidine. *Cryst. Growth Des.* **2019**, *19*, 1005–1020. [CrossRef]

287. Siddiqui, K.A. Structural Diversity of Metal–organic Hydrates: A Crystallographic Structural Database Study. *J Struct. Chem.* **2018**, *59*, 106–113. [CrossRef]

288. Infantes, L.; Fábián, L.; Motherwell, W.D.S. Organic crystal hydrates: What are the important factors for formation. *CrystEngComm* **2007**, *9*, 65–71. [CrossRef]

289. Desiraju, G.R. Hydration in organic crystals: Prediction from molecular structure. *J. Chem. Soc. Chem. Commun.* **1991**, 426–428. [CrossRef]

290. Van de Streek, J.; Motherwell, S. New software for searching the Cambridge Structural Database for solvated and unsolvated crystal structures applied to hydrates. *CrystEngComm* **2007**, *9*, 55–64. [CrossRef]

291. Bajpai, A.; Scott, H.S.; Pham, T.; Chen, K.-J.; Space, B.; Lusi, M.; Perry, M.L.; Zaworotko, M.J. Towards an understanding of the propensity for crystalline hydrate formation by molecular compounds. *IUCrJ* **2016**, *3*, 430–439. [CrossRef] [PubMed]

292. Brychczynska, M.; Davey, R.J.; Pidcock, E. A study of methanol solvates using the Cambridge structural database. *New J. Chem.* **2008**, *32*, 1754–1760. [CrossRef]

293. Brychczynska, M.; Davey, R.J.; Pidcock, E. A study of dimethylsulfoxide solvates using the Cambridge Structural Database (CSD). *CrystEngComm* **2012**, *14*, 1479–1484. [CrossRef]

294. Spiteri, L.; Baisch, U.; Vella-Zarb, L. Correlations and statistical analysis of solvent molecule hydrogen bonding—a case study of dimethyl sulfoxide (DMSO). *CrystEngComm* **2018**, *20*, 1291–1303. [CrossRef]

295. Navasardyan, M.A.; Grishanov, D.A.; Tripol'skaya, T.A.; Kuz'mina, L.G.; Prikhodchenko, P.V.; Churakov, A.V. Crystal structures of non-proteinogenic amino acid peroxosolvates: Rare example of H-bonded hydrogen peroxide chains. *CrystEngComm* **2018**, *20*, 7413–7416. [CrossRef]

296. Inokuma, Y.; Matsumura, K.; Yoshioka, S.; Fujita, M. Finding a New Crystalline Sponge from a Crystallographic Database. *Chem.—Asian J.* **2017**, *12*, 208–211. [CrossRef]

297. Byrn, S.R.; Lin, C.-T. The effect of crystal packing and defects on desolvation of hydrate crystals of caffeine and L-(-)-1,4-cyclohexadiene-1-alanine. *J. Am. Chem. Soc.* **1976**, *98*, 4004–4005. [CrossRef]

298. Leung, S.S.; Padden, B.E.; Munson, E.J.; Grant, D.J.W. Hydration and Dehydration Behavior of Aspartame Hemihydrate. *J. Pharm. Sci.* **1998**, *87*, 508–513. [CrossRef] [PubMed]

299. Saha, R.; Biswas, S.; Steele, I.M.; Dey, K.; Jana, A.D.; Kumar, S. Stabilization of 2D water sheets in a supramolecular metal–organic Schiff base complex: Reversible structural transformation upon dehydration–rehydration. *Inorg. Chim. Acta* **2013**, *399*, 200–207. [CrossRef]

300. Zhang, H.; Yin, Z. Discrete cage form of water hexamer in the hydrophilic channels assembled by heterocyclic azopyrrole. *J. Mol. Struct.* **2015**, *1092*, 9–13. [CrossRef]

301. Vologzhanina, A.V.; Zorina-Tikhonova, E.N.; Matyukhina, A.K.; Sidorov, A.A.; Dorovatovskii, P.V.; Eremenko, I.L. 36-Nuclear anionic cobalt(II) and nickel(II) complexes in solid-phase insertion reactions. *Russ. J. Coord. Chem.* **2017**, *43*, 801–806. [CrossRef]

302. Willart, J.F.; Danede, F.; De Gusseme, A.; Descamps, M.; Neves, C. Origin of the Dual Structural Transformation of Trehalose Dihydrate upon Dehydration. *J. Phys. Chem. B* **2003**, *107*, 11158–11162. [CrossRef]

303. Gillon, A.L.; Davey, R.J.; Storey, R.; Feeder, N.; Nichols, G.; Dent, G.; Apperley, D.C. Solid State Dehydration Processes: Mechanism of Water Loss from Crystalline Inosine Dihydrate. *J. Phys. Chem. B* **2005**, *109*, 5341–5347. [CrossRef] [PubMed]

304. Fucke, K.; Steed, J.W. X-ray and Neutron Diffraction in the Study of Organic Crystalline Hydrates. *Water* **2010**, *2*, 333–350. [CrossRef]

305. Braun, D.E.; Griesser, U.J. Supramolecular Organization of Nonstoichiometric Drug Hydrates: Dapsone. *Front. Chem.* **2018**, *6*, 31. [CrossRef] [PubMed]

306. Białońska, A.; Ciunik, Z.; Ilczyszyn, M.M.; Siczek, M. Discrete Cuboidal 15- and 16-Membered Water Clusters in Brucine 3.86-Hydrate, Water Release and Its Consequences. *Cryst. Growth Des.* **2014**, *14*, 6537–6541. [CrossRef]

307. Febles, M.; Pérez-Hernández, N.; Pérez, C.; Rodríguez, M.L.; Foces-Foces, C.; Roux, M.V.; Morales, E.Q.; Buntkowsky, G.; Limbach, H.-H.; Martín, J.D. Distinct Dynamic Behaviors of Water Molecules in Hydrated Pores. *J. Am. Chem. Soc.* **2006**, *128*, 10008–10009. [CrossRef]

308. CSD Python API Forum. Available online: https://www.ccdc.cam.ac.uk/forum/csd_python_api (accessed on 14 August 2019).

309. Moghadam, P.Z.; Li, A.; Wiggin, S.B.; Tao, A.; Maloney, A.G.P.; Wood, P.A.; Ward, S.C.; Fairen-Jimenez, D. Development of a Cambridge Structural Database Subset: A Collection of Metal–Organic Frameworks for Past, Present, and Future. *Chem. Mater.* **2017**, *29*, 2618–2625. [CrossRef]

310. Dolinar, B.S.; Samedov, K.; Maloney, A.G.P.; West, R.; Khrustalev, V.N.; Guzei, I.A. A chiral diamine: Practical implications of a three-stereoisomer cocrystallization. *Acta Cryst. Sect. C* **2018**, *74*, 54–61. [CrossRef] [PubMed]

311. Bryant, M.J.; Maloney, A.G.P.; Sykes, R.A. Predicting mechanical properties of crystalline materials through topological analysis. *CrystEngComm* **2018**, *20*, 2698–2704. [CrossRef]

312. Miklitz, M.; Jelfs, K.E. pywindow: Automated Structural Analysis of Molecular Pores. *J. Chem. Inf. Model.* **2018**, *58*, 2387–2391. [CrossRef] [PubMed]

313. Park, S.; Kim, B.; Choi, S.; Boyd, P.G.; Smit, B.; Kim, J. Text Mining Metal–Organic Framework Papers. *J. Chem. Inf. Model.* **2018**, *58*, 244–251. [CrossRef] [PubMed]

314. Hussain, J.; Rea, C. Computationally Efficient Algorithm to Identify Matched Molecular Pairs (MMPs) in Large Data Sets. *J. Chem. Inf. Model.* **2010**, *50*, 339–348. [CrossRef] [PubMed]

315. Zolotarev, P.N.; Arshad, M.N.; Asiri, A.M.; Al-amshany, Z.M.; Blatov, V.A. A Possible Route toward Expert Systems in Supramolecular Chemistry: 2-Periodic H-Bond Patterns in Molecular Crystals. *Cryst. Growth Des.* **2014**, *14*, 1938–1949. [CrossRef]

316. Blatova, O.A.; Asiri, A.M.; Al-amshany, Z.M.; Arshad, M.N.; Blatov, V.A. Molecular packings and specific-bonding patterns in sulfonamides. *New J. Chem.* **2014**, *38*, 4099–4106. [CrossRef]

317. Baburin, I.A.; Blatov, V.A.; Carlucci, L.; Ciani, G.; Proserpio, D.M. Interpenetrated Three-Dimensional Networks of Hydrogen-Bonded Organic Species: A Systematic Analysis of the Cambridge Structural Database. *Cryst. Growth Des.* **2008**, *8*, 519–539. [CrossRef]

318. Klamt, A. Conductor-like Screening Model for Real Solvents: A New Approach to the Quantitative Calculation of Solvation Phenomena. *J. Phys. Chem.* **1995**, *99*, 2224–2235. [CrossRef]

319. Klamt, A.; Jonas, V.; Bürger, T.; Lohrenz, J.C.W. Refinement and Parametrization of COSMO-RS. *J. Phys. Chem. A* **1998**, *102*, 5074–5085. [CrossRef]

320. Tilbury, C.J.; Chen, J.; Mattei, A.; Chen, S.; Sheikh, A.Y. Combining Theoretical and Data-Driven Approaches To Predict Drug Substance Hydrate Formation. *Cryst. Growth Des.* **2018**, *18*, 57–67. [CrossRef]
321. *Chem3D*, version 18.2 Suite; Perkin Elmer Corporation: Waltham, MA, USA, 2016.
322. CCDC Documentation and Resources. Available online: https://ccdc.cam.ac.uk/support-and-resources/ccdcresources/ (accessed on 14 August 2019).

MDPI

St. Alban-Anlage 66

4052 Basel

Switzerland

Tel. +41 61 683 77 34

Fax +41 61 302 89 18

www.mdpi.com

Crystals Editorial Office

E-mail: crystals@mdpi.com

www.mdpi.com/journal/crystals

Lightning Source UK Ltd.
Milton Keynes UK
UKHW051844160822
407383UK00002B/81